ENDE, HERMANUS VAN DEN
SEXUAL INTERACTIONS IN PLANTS:
000244159

QK827.E51

KU-405-429

WITHDRAWN FROM STOCK
The University of Liverpool

DC

1

DUE

29

DUE

7

Sexual Interactions
in Plants

EXPERIMENTAL BOTANY

An International Series of Monographs

CONSULTING EDITORS

J. F. Sutcliffe

School of Biological Sciences, University of Sussex, England

AND

P. Mahlberg

Department of Botany, Indiana University, Bloomington, Indiana, U.S.A.

Forthcoming Titles

Sexual Interactions in Plants

THE ROLE OF SPECIFIC SUBSTANCES IN SEXUAL REPRODUCTION

H. VAN DEN ENDE

Department of Plant Physiology
University of Amsterdam
The Netherlands

1976
ACADEMIC PRESS
London   New York   San Francisco

A Subsidiary of Harcourt Brace Jovanovich, Publishers

ACADEMIC PRESS INC. (LONDON) LTD.
24/28 Oval Road,
London NW1

United States Edition published by
ACADEMIC PRESS INC.
111 Fifth Avenue,
New York, New York 10003

Copyright © 1976 by
ACADEMIC PRESS INC. (LONDON) LTD.

All Rights Reserved
No part of this book may be reproduced in any form by photostat, microfilm, or any other
means, without written permission from the publishers

Library of Congress Catalog Card Number: 76 10491
ISBN: 0 12 711250 2

PRINTED IN GREAT BRITAIN BY
WILLMER BROTHERS LIMITED, BIRKENHEAD

Preface

While sexual reproduction in plants has attracted investigators through the ages, in the last 10 years there has been remarkable progress in the molecular aspects of this part of plant science. Many sexually reproducing plants, including the fungi, have been shown to be excellent systems for studying the way that cells exchange information and coordinate their activities by means of specific metabolites. In many cases, this communication is mediated over some distance by diffusible substances which are secreted into the surrounding medium by one or both of the mating partners. In several other objects, this role is played by immobile compounds, localized at the cell surface. This book is intended to serve as a starting point for students who want to investigate the molecular aspects of cell interactions in plants, and for this reason I have described a number of case histories of experimental systems which in the last few years have been shown to be amenable to this type of study. For the same reason I have paid some extra attention to the experimental aspects.

I would like to express my appreciation to my colleagues and students of the Department of Plant Physiology, University of Amsterdam, who strongly supported me during this work. In particular Drs A. Musgrave, B. A. Werkman, J. F. L. M. Urbanus, D. A. M. Mesland and R. Demets have contributed by critical discussions and reading parts of the manuscript, and permitted me to use some of their unpublished results for this book. I also thank Miss M. L. van den Briel for bibliographic help. I am most grateful to Dr A. W. Barksdale (The New York Botanical Garden), Dr H. F. Linskens (University of Nijmegen), Dr U. Näf (Manhattan College), Dr L. Machlis (University of California) and Dr R. C. Starr (Indiana University), who reviewed and

353785

commented upon parts of the manuscript. Especial thanks are due to
Dr L. Machlis, Dr E. P. Sena (University of California), Dr U. Näf
and Dr W. H. Darden (University of Alabama) for the generous pro-
vision of original photographs. I also wish to express my thanks to the
various journals and publishers, cited in the text, who have given permission to reproduce previously published material.

H. van den Ende

Contents

Et d'où savez-vous que la sensibilité est essentiellement incompatible avec la matière, vous qui ne connaissez l'essence de quoi que ce soit, ni de la matière, ni de la sensibilité? Entendez-vous mieux la nature du mouvement, son existence dans un corps, et sa communication d'un corps à un autre?

Entretien entre d'Alembert et
Diderot, 1830

1. Sexual Differentiation in Plants

Sexual reproduction is a process which recurs regularly in the life cycle of most organisms. It provides for a rearrangement of nuclear and cytoplasmic properties, in which generally, but not always, two individuals take part. The cardinal events which characterize sexuality in eukaryotes are (1) the fusion of two cells (plasmogamy); (2) the fusion of nuclei from these cells (karyogamy); and (3) reduction division of the resulting nuclei (meiosis).

The organism often prepares for these events by producing cells specialized in fusing. This occurs when its vegetative cells are not able to fuse with each other, so that plasmogamy is only exhibited by cells specialized in this respect. This implies morphological and/or physiological deviations from the vegetative state which can be quite different from one species to another. In unicellular organisms, the vegetative cell as a whole may become transformed to a sexually active cell, often without any morphological changes; in multicellular organisms, only a few cells of the whole body are specialized. They are released and live their own life for a certain period of time, or alternatively they remain part of the organism which produced them.

Fusion cells are generally equipped with or accompanied by some device to bring them into proper contact with a partner, which can be of a morphological or a physiological kind. Together with the production of fusion cells this is called *sexual differentiation*. The flowers of higher plants are illustrative of the large extent to which an organism can deviate from its vegetative growth pattern in order to effect suitable mating contact. Lower plants generally show much less elaborate sexual morphogenesis, or even completely lack it. Physiologically, however, they are equally interesting because they often exhibit, in striking simplicity, very effective mechanisms of gametic approach.

Sexual differentiation is most frequently induced or stimulated by

influences from outside, like changes in environment. Fungi are extremely
sensitive to nutrient concentration, pH, temperature, humidity and many
other factors, which all have a profound influence upon the production
of fruit bodies. This very complex subject has been reviewed by Hawker
(1957, 1966). Also in many algae the production of gametes is controlled
by nutrient supply, light conditions, CO_2 tension etc. (Coleman, 1962;
Wiese, 1969; Dubois-Tylski, 1972). Flowering in higher plants is often
dependent on the daylength: if this surpasses a critical maximal value for
a number of cycli, flowering is induced in long-day plants, whereas the
reverse is observed in short-day plants (Evans, 1969). In these examples
sexual differentiation is also induced without the presence of a mating
partner. This implies that in self-sterile plants the fertilization rate is
dependent on the synchrony of sexual development in the partners, which
is imposed by the environmental conditions. In many cases, particularly
in lower plants, a coordination of sexual development between different
individuals is realized by specific metabolites which are secreted into the
surrounding medium or are located at the cellular surface. These metabolites induce, stimulate or synchronize sexual differentiation. In some
species, to be described in the following chapters, they have been identified. In many others, their presence has been demonstrated, or inferred.

   In addition, several mechanisms have been recognized to promote
mating contact between fusion cells. *Chemotaxis* is observed in organisms
with motile gametes : they move up a concentration gradient of a metabo-
lite secreted by the generally immotile sexual partner. *Chemotropism* is
defined as growth, oriented by a concentration gradient set up by the
sexual partner. *Agglutination* is the phenomenon, found in yeasts and
many green algae, that cells of different sexual sign adhere to each other
and form massive clumps, thus increasing and maintaining cellular contact or proximity between mating partners in an efficient way.

   The developmental phenomena which promote cell fusion are subject
to mechanisms which determine the kinds of nuclei that will fuse at
fertilization. Such mechanisms are called *mating systems.* There is some
confusion as to the nomenclature of these systems, owing to the extreme
variety of sexual expression (Burnett, 1956; Coleman, 1962; Esser and
Kuenen, 1965). For the description of some sexual systems the terminology
as proposed by Kniep (1928) and Esser and Kuenen (1965) will be used
because it is more or less equally applicable to fungi, algae and higher
plants. Organisms are designated as *dioecious* when they produce either
male or female sexual structures, exclusively. Also, organisms which lack

a morphological distinction between male and female, but possess a genetically determined mating type (*plus* and *minus,* or some other deliberate indication) which accounts for a physiological rather than a morphological sex difference, are called dioecious. On the other hand, an organism is specified as *monoecious* when male and female gametes or gametangia are produced by one individual, irrespective of whether or not self-fertilization occurs.

Esser (1967) considers an organism as dioecious when it is able to operate as a donor *or* acceptor of nuclei during the sexual process; a monoecious organism, on the other hand, can be *both* donor *and* acceptor of nuclei.

In the mycological literature, the terms "homothallism" and 'hetero-thallism" are widely used since Blakeslee coined them in 1904. Generally, they denote systems with self-fertility and obligatory cross-fertility, respectively. However, the terms were originally applied to one group of fungi only (the Mucorales) and cross-conjugation is not always obligatory (see below).

A few examples may clarify the proposed terminology. The brown alga *Ectocarpus siliculosus* (cf. Chapter 9) produces motile gametes which are morphologically identical (isogametes). Some time after release, part of the gametes become immotile, and eventually are fertilized by gametes which remain motile. Therefore, the first type of gamete is designated as female and the permanently motile type as male. These types of gamete are released by two different thalli, which is reason to call this species dioecious. In the related species *Fucus serratus,* which is equally dioecious, a male/female distinction is much clearer, since the female gametes are immotile from the beginning and are much bigger than the motile male cells. In *F. spiralis* both male and female gametes are produced by the same plant. This species therefore is called monoecious.

The fungus *Mucor mucedo* (Chapter 4) produces specialized structures (*zygophores*) which fuse only with morphologically identical but genetic-ally different structures of another strain. Thus this species is termed dioecious. Since no morphological features are present to designate one strain as male and the other as female, they are labelled *plus* and *minus.* Other species related to *M. mucedo,* like *M. genevensis* and *Rhizopus sexualis,* are monoecious because sexual reproduction can occur within one individual or strain, although male and female characteristics are equally lacking.

The fungus *Achlya* (Chapter 3) is a difficult case. In most species male and female sex organs occur on the same thallus and self-fertilization seems to be most common. However, sexual interaction between two different strains is also possible, depending on the combination. It may happen that one strain behaves as a female in one type of combination and as a male in another type of combination. Nevertheless, if we apply the term monoecism to those organisms in which male and female structures are borne on the same individual, probably all species of *Achlya* are virtually monoecious.

A special case is represented by the so-called colonial algae. In the genus *Volvox*, for example, each individual consists of a colony (spheroid) of several hundreds of cells, embedded in a gelatinous envelope (Chapter 7). In every colony, male (sperm-producing) and female (egg-producing) cells may occur, or, alternatively, cells of only one sex. But also a clone of individual colonies may be completely male or female, and need another clone of different sexual sign, to reproduce sexually. The following terminology is used by Starr (1969) to distinguish between all possibilities. A *Volvox* strain is monoecious when both male and female organs arise in one and the same colony. Dioecism occurs in a strain when individual colonies produce either female or male organs. However, a clone of a dioecious strain in which only male *or* female colonies occur is denoted as *heterothallic,* while a strain which gives rise to clones in which both sexes are present is called *homothallic.* The situation can be quite complicated as is illustrated by the following example. Starr (1971) obtained four isolates of *Volvox africanus* of the following types: (a) a dioecious, heterothallic type, consisting of two strains, one producing male spheroids and the other female spheroids; (b) a dioecious, homothallic type, consisting of a single clone in which separate male and female colonies were produced; (c) a monoecious type, which formed spheroids with both male and female sexual structures; and (d) a "monoecious with males" type; clones of this type contained spheroids which were male, but also spheroids which contained male and female structures; typical females did not occur in this type.

Although in monoecious organisms self- and cross-fertilization is in principle possible, both properties are frequently restricted by a superimposed control mechanism, called *incompatibility* (Esser, 1967). One speaks of self- or cross-incompatibility according to whether restriction of self- or cross-fertility is involved. Incompatibility phenomena are genetically determined. A subtle distinction can be made between "sexual factors" (in the genetical sense), which determine sexual differentiation, including where attributable, differences between the sexes or mating types, and "incompatibility factors". Raper (1966) defines the latter as "those extrasexual genetic determinants of mating capacity which operate either in addition to or in the absence of sexual factors". Incompatibility factors thus determine mechanisms which *interfere with the sexual differentiation process prior to fertilization.* This is the reason why incompatibility phenomena can provide important information about the physiological processes underlying sexual reproduction, since they block specific steps in the normal physiological chain of reactions, or otherwise accomplish morphological deviations from the normal pattern in sexual organs which restrict fertilization (Förster, 1967).

In flowering plants, the distinction between sexuality and incompatibility is very clear, because sexual organs are easily recognized to be male or female, and the physiological stage at which an incompatibility mechanism is operative is generally well known (cf. Chapter 11). However, one runs into trouble when species are considered which are isogamous, or which lack sexual structures altogether. For instance, in fungi of the order Mucorales, where both mating partners are morphologically identical, one cannot distinguish between monoecism coupled with self-incompatibility (resulting in cross-fertility only) and real dioecism. As long as the physiological basis of sex in this group of fungi has not been elucidated, this problem remains a semantic one.

In Basidiomycetes the problem is slightly different. In these fungi no sexual organs are found prior to cell fusion. Instead, vegetative hyphae of two thalli fuse, after which an exchange of nuclei takes place. Fusion of nuclei occurs in both mycelia, unless nuclear migration and/or other stages in this process are restricted. It is generally agreed that incompatibility factors are responsible for this restrictive action. Because both mating partners deliver nuclei, in other words, are both nuclear acceptor and donor, Esser (1967) considers these organisms to be monoecists.

THE MODE OF ACTION OF HORMONES IN DIFFERENTIATION

As mentioned above, the formation of fusion cells can be induced or stimulated by the action of sex-specific metabolites, produced by the sexual partner. In general practice, such metabolites have been called *hormones.* One could define a sexual hormone as a diffusible substance playing a specific role in the sexual reproduction of the organism by which it is produced. Machlis (1972) has discussed the difficulties which are inherent in defining what a hormone is. Apart from the fact that generally the term "hormone" is used for biologically active substances which are exerting their action in the same individual in which they are produced (which in plants is not always the case), its use also implies that many unrelated compounds with widely differing actions and targets are grouped under the same heading. He therefore suggests that some generic terms might be useful to denote "classes" of sex hormones. In his terminology a sexual hormone is an "erogen" when it induces sexual structures; it is an "erotactin" when it attracts motile gametes; and an "erotropin" when it directs a growing filament. An objection which can be made against this proposal is that the classes sometimes overlap. In the fungus *Achlya*, for example, the hormone eliciting the formation of sexual structures also influences their direction of growth (see Chapter 3).

Although in many plants the induction of fusion cells has been demonstrated or suggested to occur through specific hormonal substances, only in a small number of plants have these substances been characterized (cf. Altman and Dittmer, 1973, for a compilation of the literature). Several of these identifications were accomplished at the expense of much energy and time, since biologically active compounds generally are produced in very low amounts, their objects having a correspondingly high sensitivity. The endeavour to elucidate the structure of low molecular-weight sex hormones, or to prepare high molecular-weight substances in absolute homogeneity, is nevertheless most valuable because only then can sensible studies be made about their mode of action, their target and metabolism.

Prerequisite to any characterization of a biologically active substance is the need to develop a reliable method to assay its presence and concentration, because chemical or physical methods usually lack sufficient sensitivity or specificity, or both. When a quantifiable response is obtained, the construction of dose-response curves are usually required for an accurate determination of hormone concentration. In addition, a curve obtained with a standard sample is normally required. Because of biological variability, a standard curve once determined may not apply at some other time, and must therefore be included in each assay. Unfortunately, standard samples are not always available, owing to poor stability of the compound in question. For that reason the concentration is often expressed not in terms of a certain amount of standard preparation but in terms of the lowest amount giving a detectable response. Of course, the accuracy is then directly dependent on the dilution factors used, and again, is subject to the variability of the test organism. Finney's book (1964) may be consulted for more information about the theory and methodology of biological assay.

As will become apparent from the following chapters, very little experimental material is available about the mode of action of substances inducing sexual reproduction. Only a few reports are present of early biochemical changes following administration of sex hormones to hormone-deprived plant cells; equally rare are investigations concerned with the fate of a hormone as it exerts an effect. This is because only in the last few years have some of the known hormones become available in pure form. We know much more about other hormones, particularly in the animal kingdom. From these studies the idea has evolved that cells which are particularly sensitive towards a certain hormone, called "target cells", contain receptor molecules to which the hormone specifically binds. The complex of hormone and receptor molecules triggers secondary reactions, eventually affecting metabolism, morphogenesis etc. The affinity of the hormone for its target cell is in the first place determined by the association constant of the hormone–receptor complex and by the number and accessibility of the receptor molecules.

A few examples of intensively studied hormonal actions in animals and plants may be given for comparison with the material that will be described in the following chapters.

Several peptide and catecholamine hormones in mammalian systems appear to function with adenosine-3', 5'-monophosphate (cAMP) as intermediary. In the hormonal effect a direct interaction between the hormone and the enzymes adenyl cyclase or cAMP diesterase is involved. Adenyl cyclase catalyses the formation of cAMP and pyrophosphate from ATP. cAMP diesterase stimulates the conversion of cAMP to 5'-AMP. Consequently the hormonal action leads to a change of cAMP concentration which may lead to a multitude of effects, dependent on the type of cell which is involved (Butcher and Sutherland, 1962).

In a number of cases it has been demonstrated that the hormonal action on adenyl cyclase activity proceeds via specific receptor molecules. In mammalian fat cells only one adenyl cyclase, located in the plasma membrane, is stimulated by different hormones such as glucagon, adrenocorticotropin and epinephrine; this action is mediated by hormone-specific receptors located at the outer surface of the membrane (Rodbell *et al.*, 1970; Cuatrecasas, 1974).

The most pronounced action of cAMP in eukaryotes is its ability to activate or inactivate enzymes that catalyse key reactions, particularly those concerned with the mobilization of potential reserves of carbon and energy (reviewed by Jost and Rickenberg, 1971). The most thoroughly investigated example of cAMP action is that on the degradation of glycogen in the liver. In this case it acts in response to the effect of the hormone epinephrine on liver-cell adenyl cyclase (Villar Palasi and Larner, 1970).

cAMP has also been demonstrated to be functional in several plants, like *Chlamydomonas* (Amrhein and Filner, 1973), *Mucor* (Larsen and Sypherd, 1974), *Coprinus* (Uno and Ishikawa, 1973), *Neurospora* (Terenzi *et al.*, 1974) and many others. It is uncertain, however, whether this compound has an universal function in plants as it has in animal systems.

Characteristic of this and other systems which imply the activation or inactivation of enzyme molecules that are already present is that the hormone action is very rapid. Epinephrine action, for instance, has a lag period of a few minutes. On the other hand, the administration of hormones influencing protein synthesis evokes an effect only after a much longer period of time. This is illustrated by the action of steroid hormones, involved in mammalian sexual differentiation. These also interact with specific receptor sites, present in target cells. Work done on the primary action of estradiol, reviewed by Jensen and De Sombre (1972), indicates that this compound is accumulated in cells of the uterus against a concentration gradient. This is caused by the formation of a complex between the hormone and specific receptor proteins which are particularly abundant in the cytoplasm of uterine cells. The association leads to a change in properties of the protein, which is reflected by a change in sedimentation velocity. The complex is transported to secondary receptor sites in the nucleus which again are characteristic of the target tissue. The association of the complex with these receptor sites evokes more or less specific reactions leading to increased uterine growth. These involve enhanced capacity for RNA synthesis, 1–2 h after hormone administration, followed by an increased rate of protein synthesis.

Progesterone and other steroid hormones also appear to be subject to a two-step mechanism in which a complex is formed between the hormone and a cytoplasmic receptor protein which consecutively is transported to a second receptor site in the nucleus (Jensen *et al.*, 1971). Although the cytoplasmic protein may just function as a carrier to guide the hormone molecules in maximal numbers to their site of action, it is also possible

that it is the chief agent inducing some nuclear process, the function of
the steroid being to promote its transformation to an active form which
can reach and bind to the nuclear acceptor site (Jensen and De Sombre,
1972).

Turning to the plant kingdom, a number of growth regulatory substances can be discussed which play an important role in plant morphogenesis (reviewed by Steward and Krikorian, 1971; cf. also Carr, 1972;
Kaldewey and Vardar, 1972). Three examples will be mentioned.

The *auxins* are a class of compounds exemplified by indoleacetic acid,
the major action of which is to promote cell elongation. Since plant cell
walls are surrounded by a moderately rigid cell wall, one of the major
aspects of indoleacetic acid action is to stimulate cell wall extension.
Much work has been devoted to the effects of this hormone on RNA and
protein synthesis (reviewed by Ray, 1974). However, these studies do not
account for the remarkable property of this hormone to elicit a growth
response within 10 to 15 min. Such a rapid effect appears to be correlated
with the activation of glucan synthetase and xyloglucan metabolism in the
cell wall (Ray, 1973; Labawitch and Ray, 1974). Another fast action of
indoleacetic acid is the stimulation of cellular K^+ and Cl^- uptake and
excretion of H^+ ions through the plasma membrane (Rubinstein and
Light, 1973; Cleland, 1973). The latter effect is considered particularly
interesting because it could decrease the pH in the cell wall, which would
extend passively in response. This might result in the onset of cell enlarge-
ment, in the long term supported by stimulated cell metabolism (cf. review
by Evans, 1974). These studies have led to the view that the primary
action of indoleacetic acid is in the plasma membrane. This is strengthened by the fact that a specific binding has been demonstrated of
indoleacetic acid to membrane fragments of homogenates of maize
coleoptiles (Hertel *et al.,* 1972). Analogues of indoleacetic acid that are
active in growth stimulation can substitute for indoleacetic acid at the
presumed binding sites, whereas chemically similar but biologically
inactive molecules do not compete in the binding. The nature of the
receptor molecules has not been clarified.

The multiplicity of actions exhibited by indoleacetic acid (involving
ion transport, RNA and protein synthesis, activation of enzymes) is also
observed with other higher plant hormones. As a second striking example
the *gibberellins* can be mentioned (reviewed by Jones, 1973). Their most
pronounced effect is to elongate the stem and leaf sheaths in a large
number of plants. The difference between genetically dwarf and tall plants

may reflect deficient gibberellin production in the dwarf plants. Frequently gibberellins and auxins act synergistically. But gibberellins have many other functions in plant growth and differentiation. They promote bolting and flowering in several types of long-day plants, influence sex expression, induce parthenocarpic fruit set (i.e. fruit development without prior fertilization), and are involved in the breaking of dormancy, seed germination and cell division. Only one phenomenon has been analysed in detail, namely the induction of starch hydrolysis in barley seeds. Gibberellins, produced by the embryo of the seed, induce the production of α-amylase in the layer of aleurone cells, surrounding the endosperm (reviewed by Varner, 1974). These aleurone cells are the "target cells" that respond to gibberellin. This system thus provides an example of organ specificity in hormone action which is rare in higher plants.

The mechanism by which α-amylase synthesis and secretion is regulated, is unknown. A relatively rapid action of gibberellin (with a lag period of approximately 2 h) which may have a bearing on the regulation of enzyme synthesis is the massive proliferation of rough endoplasmic reticulum (Jones, 1969) and the increase of the activity of phosphorylcholine-cytidyl and phosphorylcholine-glyceride transferases which are involved in lecithin biosynthesis (Johnson and Kende, 1971; Ben-Tal and Varner, 1974). As suggested by Johnson and Kende, these enzymes might be at least partly responsible for the hormonally stimulated membrane synthesis. It is well established that the rough endoplasmic reticulum participates in the synthesis and exportation of secreted proteins in eukaryotic cells. The gibberellin-induced synthesis of hydrolytic enzymes in aleurone cells might take place only on polysomes that are bound to newly synthesized endoplasmic reticulum, cytoplasmic proteins being produced on free ribosomes. This is a kind of control of the translation of existing messenger RNA (mRNA) species. Another possibility to be envisaged is that the rate of secretion of hydrolytic enzymes (controlled by the rate of membrane production) is determining the rate of synthesis of these enzymes. An analogous situation is found in *Achlya,* where the rate of cellulase production is limited by the rate of its secretion (Thomas, 1974; cf. Chapter 3).

The action of gibberellins on the synthesis of endoplasmic reticulum is reversed by *abscisic acid,* a hormone structurally related to trisporic acid (see Chapter 4). Some other effects of abscisic acid have been shown to occur within a very short lag period: a regulation of stomatal aperture (Cummins *et al.,* 1971) and a change of plasma membrane potential

(Tanada, 1972). This suggests that abscisic acid has an action on plasma membrane properties.

THE ROLE OF THE CELL SURFACE IN DIFFERENTIATION

As will be shown in subsequent chapters, there is evidence that in some cases metabolic changes which are part of sexual differentiation are triggered by cellular contact. For example, the formation of gametangia in many fungi only occurs after cellular contact with the mate has been established. In *Hansenula,* cellular contact seems to be pre-requisite before conjugation tubes are formed. Also flagellar adhesion between gametes of green algae is an outstanding example illustrating the importance of cellular contact in sexual differentiation. The inductive role attributed to the cell surface in these and similar cases is in line with evidence emerging from studies with animal cells. The morphogenetic interaction between cells during embryonic development and the wide range of immunological reactions illustrate the importance of the cell surface in intercellular relations (Martz and Steinberg, 1972; Merrell and Glaser, 1973; Balsamo and Lilien, 1974; Feldman and Globerson, 1974). This implies that a high degree of molecular specificity must be present at the cell surface to explain these phenomena. This also holds for plant cells, even while they are surrounded by a relatively thick cell wall.

Although the importance of the cell surface in morphogenesis has long been recognized, analytical studies to explain its role on a molecular level have practically been restricted to animal systems. It seems appropriate, in this introductory review, to give a short account of some prevailing hypotheses.

The first suggestion, which has been popular for a very long time, is the hypothesis published by Weiss (1947) and Tyler (1947). It is based on the assumption that cell surfaces contain substances which interact specifically like antigens and antibodies, resulting in cellular adhesion and evoking secondary responses. In fact, cells are known to carry antigenic determinants at their surface, such as bacterial antigens, blood group substances and histocompatibility factors, while bone-derived lymphocytes contain antibody-like immunoglobulins as specific receptor sites of antigens which may be cellularly bound (Feldman and Globerson, 1974). The interaction between antigen and antibody is of a non-covalent nature and is governed by spatial characteristics, and although not demonstrated experimentally, it is reasonable to assume that the association of these two molecular species leads to a conformational change in the globulin receptor, which might be functional in eliciting cellular responses such as antibody production in lymphocytes and phagocytosis in macrophages (Metzger, 1970; Allison *et al.*, 1971). On the basis of this model the interaction between two cells could imply a conformational change in surface receptors of one cell, caused by ligand molecules present on the surface of the other cell. In the case of interaction between identical cells, both reacting elements would be present on the same surface but be spatially separated. The property of cells to recognize each other would thus be based on the specificity of interaction between the two types of surface elements.

An outstanding example of cellular interactions which is reminiscent of antigen–antibody interactions is aggregation in sponge cells (reviewed by Kuhns *et al.*, 1973). This aggregation, which precedes differentiation and growth of the organism, is mediated by a high molecular-weight glycoprotein, released from the sponge cell surface. This glycoprotein, called "aggregation factor", contains specific carbohydrates as determinants of species specificity (Burger *et al.*, 1971). This is suggested by the fact that glucuronic acid is an effective inhibitor of aggregation in *Microciona prolifera*. This inhibition is only observed when chemically dissociated cells are preincubated with glucuronic acid before the aggregation factor is added. Glucuronic acid has no effect on the aggregation of other species, however (Turner and Burger, 1973).

A number of cell surface interactions are only temporary. Yeast cells adhere only prior to and during the conjugation process. The fused cells have lost their adhesive properties (see Chapter 5). In *Chlamydomonas*, *plus* and *minus* gametes adhere to each other with their respective flagella. Again, as soon as the fusion is complete, the flagella lose their adhesive properties (Chapter 6). As pointed out by Roseman (1970), in animal systems also specific adhesions may be temporary. He therefore proposed a hypothesis which has met considerable acclaim. The mechanism he proposes is based on the enzyme–substrate model. Cell surfaces contain both substrates and enzymes, and the binding of one to the other results in adhesion. On completion of the enzymatic reaction the cells may dissociate again. This model further assumes that complex carbohydrates and glycosyltransferases are involved in the adhesion. Generally, surface carbohydrates of animal cells appear to consist of glycoproteins and glycolipids which are an integral part of the plasma membrane (cf. Nicolson, 1974). Recent evidence indicates that glycosyl transferases are functional in the biosynthesis of these complex carbohydrates which transfer sugar moieties from sugar nucleotides to the end of a sugar chain of a glyco- or lipoprotein. An important aspect of many of these enzymes is their specificity for the acceptor molecule. The product of one reaction becomes the substrate for the next enzyme in the sequence of reactions leading to the completion of an oligosaccharide chain. Thus classes or families of enzymes might exist, like galactosyl transferases and N-acetylglucosaminyl transferases in which each member utilizes the same sugar donor but shows a different specificity with regard to the acceptor molecule (Roseman, 1974). These enzymes are primarily located in intracellular organelles (like the Golgi apparatus) but they have also been observed to be present at the cell surface (reviewed by Roth, 1973). According to Roseman's hypothesis, these surface-located glycosyl transferases may not only be involved in the synthesis of surface carbohydrates, but may also be able to bind to an appropriate sugar acceptor on the surface of a neighbouring cell. This would lead to the association of the two cells. In the absence of a sugar nucleotide as sugar donor this association would have a permanent character (analogous to the association of hexokinase and glucose in the absence of ATP; Roseman, 1974), whereas the availability of the sugar donor would ultimately lead to the dissociation of the surface-bound enzyme and the product of the reaction bound to the neighbouring cell, and thus to the dissociation of the two cells. Therefore, the availability of the sugar donor might exert considerable control over the adhesive process. The observed specificities of glycosyl transferases seem sufficiently high to explain the specificities encountered in cell-to-cell surface interactions and to attribute an element of recognizability to cell surfaces which contain these complex carbohydrates.

Suggestive evidence for this hypothesis has been summarized by Roth (1973) and Roseman (1974). Some intact cells carry out transferase reactions with exogenous acceptors if they are incubated with sugar nucleotides in the surrounding medium. This transfer must occur at the cell surface because the sugar donor does not enter the cell. Radioautographic analysis of cells after incubation with a labelled sugar nucleotide shows that the acceptor is localized at the periphery of the cell. Normal mouse embryonic cells which exhibit inhibition of growth upon cell contact contain galactosyl transferases at the outer surface which transfer galactose from uridine diphosphate galactose to acceptors of adjacent cells only after intercellular contact is made (called *trans*-glycosylation). Malignant cells, on the other hand, which show relatively much less contact-inhibition of growth, are able to catalyse the transfer of galactose to acceptors located on the same cell (*cis*-glycosylation; Roth and White, 1972). Specific adhesion between neural retinal cells of chick embryos can be perturbed by carbohydrates that also act as galactosyl acceptors for the cell-surface transferases on these cells (Roth *et al.*, 1971).

What is the relevance of these considerations for plant cells which naturally are surrounded by a relatively thick cell wall? The function of the wall seems to be in the first place to confer a degree of rigidity upon the plant cell, and to provide it with a microenvironment which is relatively well balanced with respect to water and ion content. One could argue that besides these beneficial effects a thick cell wall would prevent the cell from communicating with adjacent cells by means other than diffusible hormonal substances. Nevertheless, there is good reason to assume that plant cells which are in contact can also communicate without diffusible factors. There are many phenomena in higher plant morphogenesis, like the development of the stomatal complex and the symmetry of secondary wall growth leading to pits or rings in vascular tissue, which suggest an intercellular communication system in which the cell surface plays a dominant role (Newcomb, 1969). The effects of cellular contact which are observed during sexual reproduction in plants might in particular be illustrative of the importance of cell-surface properties and might serve as experimental material to test the ideas described above.

THE PROMOTION OF SEXUAL CONTACTS

In many plants cellular contact between compatible fusion cells is promoted by chemotaxis, chemotropism, agglutination and gel trapping. The last phenomenon has been observed in a number of algae, like *Oedogonium* (cf. Chapter 8), where the egg cell secretes a mass of mucilage or slime in which free-moving male gametes are trapped. Thus these gametes are immobilized in the neighbourhood of the egg cell which evidently results in an increased fertilization rate. Also, paired gametes of desmids wrap themselves in mucilage, which presumably holds them together in anticipation of the conjugation process. However, this mechanism of bringing or keeping gametes together has not been much investigated in plant cells.

Another means of promoting sexual contact in unicellular organisms is agglutination, which can be defined as a mass adhesion between sexually competent cells. In all the cases which have been studied, this type of cell adhesion shows very high specificity with regard to sex and species. It is important to note that this type of interaction does not necessarily result in the transfer of information from cell to cell, as is characteristic of the inductive cellular contact which was discussed in the preceding section. Particular attention will be paid to this subject in Chapter 5.

Chemotaxis

Among the chemical stimuli operative in plants, the one resulting in chemotaxis of motile cells has attracted considerable attention. The older literature has been reviewed by Machlis and Rawitscher-Kunkel (1967). They emphasize that chemotaxis is usually not observed in sexual systems where both sexual partners are equally motile. Normally, chemotaxis is only observed in species where gametes of both sexes exhibit considerable difference in locomotive activity, or where one of both is completely sessile. Only then can a concentration gradient be effective as an orienting vector.

Pfeffer (1884) initiated and established the field of study concerning directed movements. He noted in a large number of organisms that the following three parameters are important: the nature of the attracting or repelling substance, the form of the concentration gradient and the absolute concentration of this substance. We will return to these aspects below when dealing with bacterial chemotaxis.

One other more or less general characteristic, noted by Machlis and Rawitscher-Kunkel (1967) and Wiese (1969), may be mentioned. Chemotaxis is often followed by attachment of the motile cell to the cell emitting the attractant. This has led Wiese to speculate that chemotaxis and surface adhesion might be coupled phenomena in these cases : "A functional surface structure may be envisaged as responsible for the last step of chemotaxis and for the initial agglutinative fusion step. Just as the chemotactic substance is supposed to interact with some absorbing receptor on the androgamete, one may visualize a similar interaction of these receptors with the emitting surface of the gynogamete. No such case is sufficiently investigated." (Wiese, 1969.)

The relationship between attraction and adhesion has been particularly emphasized by Held (1974), in a study of the host–parasite relationship between the fungi *Rozella allomycis* and *Allomyces arbuscula*. The host genus, *Allomyces*, secretes a diffusible factor which attracts zoospores of *Rozella*. Moreover, its cell surface contains elements ("receptors") which

cause zoospores to adhere and to encyst. Both the attraction and attachment reactions are quite specific for zoospores of *Rozella*. The related
species *Blastocladiella emersonii* also attracts *Rozella* zoospores but supports very limited attachment. Evidently, both phenomena can be un-
coupled to certain extent. When zoospores are attracted via a dialysis
membrane, they do not attach to the surface of the membrane. Neither
does *Rozella* settle on dead hosts, which presumably no longer exude the
attractant. Random collision with the host surface is not sufficient to
result in attachment. These observations suggest that in this host–parasite
relationship attachment is dependent on direct contact with the host
surface and also on the presence of attractant. How the attractant is
involved in the adhesion reaction is not known. This example is very
similar to situations where gametes are attracted by and attach to egg
cells, as in *Fucus* or *Allomyces*, and also to those where taxis seems to
precede sexual agglutination, as has been described for species of
Chlamydomonas (Tsubo, 1961) and *Pandorina* (Rayburn and Starr,
1974).

   In the last ten years considerable progress has been made in the field
of bacterial chemotaxis. In particular the effects of a concentration
gradient on motility and the problem of gradient-sensing by a small
organism are now better understood. Because several problems are common to both bacterial and eukaryotic types of chemotaxis, and because

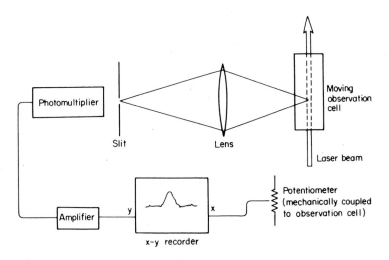

FIG. 1. A schematic representation of the apparatus used to monitor bacterial
chemotaxis by light scattering (from Dahlquist *et al.*, 1972).

it is reasonable to assume that the chemosensory mechanism may be at least formally analogous in both types of organisms, some reference to studies of bacterial chemotaxis may be useful. In addition, some of the techniques used to measure taxis quantitatively might be applicable to bacteria and gametes alike.

Such a technique has been described by Dahlquist and coworkers (1972) and is depicted in Fig. 1. Various distributions of bacteria and/or attractants are made in an oblong cell, 10 ml volume and 8 cm long. For example, a homogeneous distribution of bacteria and an exponential, one-dimensional gradient of serine as attractant may be combined. The gradient is stabilized with glycerol. The density of the bacterial suspension at various positions in the cell is determined by monitoring along the length of the cell the intensity of light scattered by the bacteria. The light is supplied by a He-Ne laser which shines up through the bottom of the cell parallel to its long axis. A photomultiplier tube measures the intensity of the laser light scattered at right angles to the beam. The distribution of bacteria can be scanned in 1 to 3 min and recorded automatically, and this distribution can be followed as a function of time. Figure 2 shows an

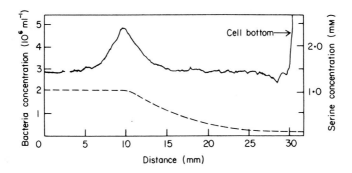

Fig. 2. The response of *Salmonella typhimurium* to an exponential gradient of L-serine. The trace represents the concentration of bacteria after 15 min. The dashed line represents the serine concentration (from Dahlquist *et al.*, 1972).

example of the distribution of bacteria to an exponential gradient of serine. The results of experiments of this type were interpreted in terms of Weber's law, which states that the magnitude of the chemotactic response depends on proportional changes in concentration of the attractant. In other words, not the absolute gradient dc/dx but the proportional gradient

$$\frac{\mathrm{d}c/c}{\mathrm{d}x} = \frac{\mathrm{d}\ln x}{\mathrm{d}x}$$

determines the average velocity of the bacteria. This result confirms earlier statements made by Pfeffer (1884). Figure 2 suggests that bacteria are moving through an exponential gradient at a steady rate. However, this relationship holds only over a narrow concentration range because the average velocity of the bacteria appears also to be dependent on the absolute concentration gradient over a range of several orders of magnitude. There is a clear optimum at 10^{-4}–10^{-2}M with respect to tactical motility. Probably at higher concentrations the gradient-sensing system in these bacteria is less operative because it has become saturated. Other quantitative techniques have been described by Nossal and Chen (1973) and Mesibov *et al.* (1973).

How is the motility of bacteria influenced by attractant gradients? Koshland (1974) reviews evidence that *Salmonella typhimurium* and *Escherichia coli* (the two most investigated species) move in straight lines and turn abruptly as a consequence of tumbling. The lengths of the straight runs and the angles of the turns have a random distribution. From quantitative analysis it appears that the average length of a run is increased when a bacterium travels up a gradient of attractant. This results in an approach towards the source of the attractant, even while the movements of the bacterium remain essentially fortuitous ("biased random walk"; Berg and Brown, 1972; Koshland, 1974). A change in the direction of flagellar rotation appears to be the basis of the chemotactic response in *E. coli* (Larsen *et al.*, 1974).

The next problem, how bacteria are able to detect a gradient, has been elegantly resolved by Macnab and Koshland (1972), who reasoned that bacteria might be too small to detect spatial differences in the concentration of an attractive substance. The ability to make temporal comparisons between concentrations seemed much more plausible in a relatively fast-moving organism. They tested this idea by developing a "temporal gradient apparatus". Bacteria, initially present in an uniform attractant concentration, were plunged by a rapid mixing device into another (higher or lower) uniform concentration. They were then immediately observed by microscopic and photographic techniques. The idea was that on complete mixing, the bacteria would respond as if they were in a uniform environment in the case of spatial gradient sensing, but would respond as if they were in a gradient when sensing a concentration change as a function of time. It appeared that bacteria subjected to a decrease in concentration of attractant tumbled more frequently and changed direction consequently more often than normal immediately after mixing; when subjected to an increase in concentration, bacteria tumbled far less frequently and made longer runs. It can be concluded from this result that a time-based process is operative in chemotaxis of bacteria; in other words, they have some sort of "memory". When moving up a gradient a bacterium notes that the concentration at time t is higher than at time 0, which tells it to tumble less frequently. This results in a net movement in the direction of the highest concentration.

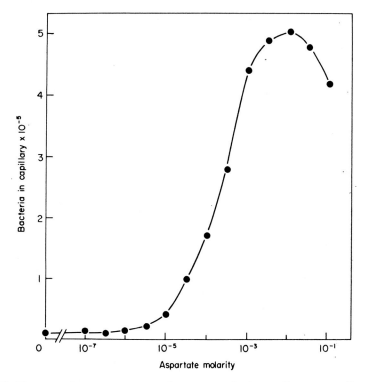

FIG. 3. Concentration–response curve for aspartate in the capillary assay of bacterial chemotaxis (from Adler, 1973).

The receptor system that detects the chemotactic stimulus was eluci-dated to large extent by Adler and coworkers. He used the following quantitative assay for measuring chemotaxis (Adler, 1973). A capillary containing a solution of attractant was placed in a suspension of bacteria, under conditions allowing motility and taxis but no growth. After incubation at 30° for 1 h the capillary was removed and the number of bacteria inside the capillary determined by plating its contents and counting colonies the next day. A typical concentration–response curve is shown in Fig. 3. The lowest concentration of attractant that produces an accumulation of bacteria in the capillary appeared to be in the first place dependent on the chemotactical activity of the solute tested but also on the density of the bacterial suspension. This is because the attractant is generally metabolized. This results in a decrease of the effective attractant concentration. When non-metabolizable attractants are used, or mutants that cannot metabolize or absorb the attractant, the lower threshold is not affected by the bacterial density. The concentration which gives maximal response is dependent on the nature of the attractant and, again, on the rate of metabolism of the attractant. At very high concentrations so much attractant diffuses out of the capillary that the bacteria are saturated before entering it.

With this technique a variety of sugars and sugar derivatives were tested for their chemotactic activity. At least nineteen of them appeared to be active at low threshold concentrations. By competition experiments and the use of numerous mutants it was established that nine different chemo-receptors are involved in the detection of these various attractants, with different specificities (Adler *et al.*, 1973).

A galactose-binding protein was isolated from cells by mild osmotic shock which appeared not to be present in some of the mutants unable to respond tactically to galactose, or others which were defective in galactose uptake. Removal of this factor from wild-type bacteria virtually eliminated the capacity to respond tactically towards galactose. This could be restored by the addition of a solution of purified factor (Hazelbauer and Adler, 1971). This, and other evidence (reviewed by Koshland, 1974, and Berg, 1975) has led to the conclusion that the galactose-binding protein is part of the galactose transport system of the attractant through the plasma membrane, although transport is not required for chemotaxis. Mutants are known which are normal in taxis and defective in the uptake of a certain compound. The reverse has also been observed: mutants which have normal transport ability but do not react tactically. This indicates that more functions are involved in gradient-sensing than binding alone. The chemoreceptor probably is primarily responsible for sensing the presence and the concentration of attractant and transmitting this informa-tion to some other functional part of the cell, including the motor apparatus.

Several other proteins have been isolated with similar characteristics, with specificity for other attractants (cf. Koshland, 1974).

These studies show that unicellular organisms provide useful model systems for the study of sensory perception. Chemotaxis has also been studied in leucocytes (Wilkinson, 1974), cellular slime moulds (Bonner, 1973; Gerisch *et al.*, 1974) and insects (Karlson, 1973).

Chemotropism

Chemotropism, the directional response of a growing filament to the concentration gradient of a particular substance, is a widespread phenomenon among fungi. It is quite uncommon, however, among algae and higher plants. In these organisms only copulation tubes (as in *Spirogyra*), pollen tubes and root hairs are presumably chemotropically oriented. Pollen tube growth is the subject of Chapter 11, in which the evidence for chemotropism in this system is described. In this section the discussion will be restricted to some general remarks on chemotropism in fungi.

There is very little information about directed growth of fungal hyphae. The older work has been reviewed by Ziegler (1962). Some authors have claimed that chemotropism plays an important role in vegetative growth of fungi. Fischer and Werner (1955), for example, have demonstrated that the hypha of *Saprolegnia* species grow towards a source of amino acids. This response might be of ecological significance, since amino acids are good nutrients for this type of fungi.

Some attention has also been paid to the opposite phenomenon, namely *autotropism*. This is negative chemotropism with respect to other hyphae of the same individual or species. In a spore suspension the point of the germ tube from a spore is determined to large extent by the location of neighbouring spores, and so is the direction of germ-tube growth. Also, neighbouring somatic hyphae generally avoid each other which results in the characteristic growth pattern of a fungal colony on a solid medium. Conceivably, this behaviour is caused by the continuous production by hyphae of substances which have a negative tropic effect on neighbouring hyphae. Robinson (1973a) has reviewed the evidence. Such substances have never been isolated, however. Robinson (1973b) has proposed that the autotropic responses are due to positive chemotropism to oxygen, in an environment where oxygen is at a low concentration and is the factor most likely to limit hyphal growth.

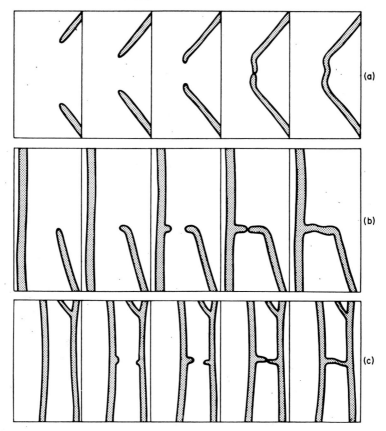

Fig. 4. Diagram showing successive stages in (a) hypha-to-hypha, (b) hypha-to-peg, and (c) peg-to-peg fusions in fungi (from Buller, 1933).

A well-established fact is the orientation of growth preceding somatic cell fusion in Ascomycetes, Basidiomycetes and Fungi Imperfecti. As a consequence of vegetative fusions (anastomoses) between neighbouring hyphae a mycelium may become a three-dimensional network. Buller (1933) remarks that it is characteristic of those fusions to be preceded by an "action at a distance", and he gives many examples from the older literature. According to this author, this action at a distance may be of two kinds: an action by one hypha causing another hypha to send out a side branch and an action which causes the first hypha and the branch to grow towards one another until they meet. These two actions result in three types of hyphal fusions, which are frequently encountered together in a colony: hypha-to-hypha, hypha-to-peg and peg-to-peg fusions. These

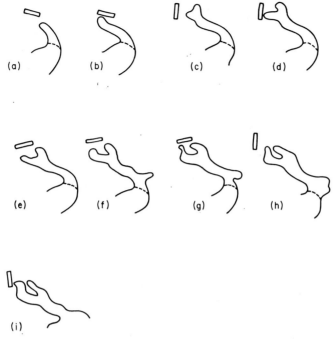

Fig. 5. Chemotropism of the trichogyne in *Ascobolus stercorarius*. (a) 6.00 p.m., oidium and ascogonium; (b) 6.12 p.m., directional growth of the trichogyne; (c) 6.25 p.m., development of a lateral branch as a response to a relocation of the oidium made at 6.16 p.m.; (d) 6.37 p.m., growth of the branch directly to the oidium and a change in direction of growth of the trichogyne tip; (e) 6.41 p.m., oidium moved to new position; (f) 6.48 p.m., both apices responded to the relocation of the oidium; (g) 6.53 p.m., both apices grew directly to the oidium; (h) 6.57 p.m., oidium moved a third time; (i) 7.21 p.m., only one of the apices responded to the relocation (redrawn from Bistis, 1957).

are illustrated in Fig. 4. In hypha-to-hypha fusions two filaments grow into each other's influence. Because of the fact that hyphae extend only at the tip, any new cell material becomes oriented towards the other hypha, whereas what was present before remains where it was. As shown, this type of growth leads to very sharp curvatures. In the second type of fusion, a secondary growth tip is initiated as the result of the proximity of a hyphal tip which is attracted to an older part of a hypha. Next, this hypha and the newly formed one grow towards each other (Fig. 4b). In the third type, bridges are formed by the initiation of new tips at either hypha. This type of fusion is especially found between hyphae lying very close to each other (0–5 μm; Fig. 4c).

It appears from Buller's work that reorientation of hyphal growth is often accompanied by the formation of side branches. This is also, and most spectacularly, observed in *Achlya,* a fungus in which hyphal growth is probably directed by a well-defined hormone, antheridiol (Chapter 3, cf. Fig. 9). With respect to another well-known chemotropic system, *Ascobolus stercorarius,* Bistis (1957) has stated that there is essentially no difference between growth orientation and branching. In this fungus a sexually differentiated hypha, the trichogyne, is chemotropically directed towards a compatible filament with which it eventually fuses. This system seems particularly amenable for the study of chemotropism because the directing cell can be very small, for example an oidium or a hyphal fragment, which in first approximation can be considered as point sources of a chemotropic agent. In addition such sources can be repositioned to study the concomitant changes in the hypha directed towards it (Bistis, 1956, 1957). It is most unfortunate, therefore, that the nature of the responsible agent has not been clarified.

An examination of some of the photographs presented by Bistis and schematically redrawn in Fig. 5 leads to the following suggestions: (1) the change in direction of growth of a trichogyne in response to a relocation of the sexual partner is caused by the action of an agent diffusing from the sexual partner; (2) the change in the direction of growth of the trichogyne in response to this relocation is not due to a curvature but rather to the establishment of a new growing point. This new growing point is usually subapical and occurs on the side nearest to the relocated partner (Bistis, 1957).

A second phenomenon which sometimes accompanies oriented growth is the *stimulation* of growth. This is particularly apparent in pollen tubes (see Chapter 11) but also in fungi. In *Mucor mucedo* sexual hyphae of different mating types have an attractive action on one another which not only results in reorientation but also in stimulation of growth (Plempel, 1962; Mesland *et al.,* 1974; see page 72).

Apart from a few general observations as mentioned above, there is very little known of chemotropism. A number of circumstances may account for this. In the first place there is very limited knowledge about the regulation of apical growth (cf. a recent review by Bartnicki-Garcia, 1973). As long as it is not known by what factors cellular extension is restricted to a very limited zone at the apex, and by what factors the rate of this extension is determined, it is hard to predict how this extension can be modulated to result in reorientation of growth. Secondly, to study a phenomenon like this requires a reliable assay procedure, which among others discriminates between growth orientation and mere growth stimulation (see also page 155). Such a procedure has not been available up to now. Some methods designed to detect chemotropism of pollen tubes will be considered in Chapter 11. Three other techniques have been reviewed by Robinson (1973a), the direct method, the capillary method, and the perforated plate method. In the direct method agar blocks containing the test material are pushed against or positioned upon a solid medium on which a mycelium is growing. Any resultant tropism is noted by visual inspection. This technique has been used by Fischer and Werner (1955) to observe positive orientation of the hyphae of *Saprolegnia* to a source of amino acids. In the capillary method, a capillary filled with a solution of the test compound is brought into contact with its open end in the neighbourhood of a mycelium. This qualitative method is based on the well-known method of studying chemotaxis in free-moving organisms (cf. the preceding section). In the perforated-plate technique a mycelium is grown in the vicinity of holes in a perforated sheet which separates the solid substrate from an opposite layer of agar containing the test material. This method was used by Robinson (1973a) to study autotropism in germ tubes of *Geotrichum candidum*. All these methods depend on direct observation, with the risk of prejudiced judgement, and are semiquantitative at best.

An interesting technique for the quantitative study of chemotropism was developed by A. Musgrave (unpublished work) for germ tubes of *Achlya*. Zoospores of this organism are spread evenly over the surface of plain agar strips (11 × 1 × 0·15 cm). Then a 1 cm² block is removed from the centre of the strip and another block containing the test compound (donor block) is put in its place. The agar strips are then pushed into contact with the donor block so that the test material can diffuse into the strips. Twenty hours later most of the spores have germinated and where the hyphae are within a gradient of a compound that induces chemotropism they grow along the strip towards the donor block. The chemotropic growth of hundreds of germ hyphae cannot be confused with non-chemotropic growth, and so the line of demarcation where chemotropic growth ends and random growth begins can be determined with accuracy. When the distance of these lines from the donor block are plotted against the logarithm of the original concentration at the source, a linear relationship is observed from which one can determine the lowest concentration that supports chemotropism. At high concentrations, hyphae

B

close to the donor block branch frequently and show no sign of growing towards the block. Thus chemotropism is lost at higher concentrations and again lines of demarcation can be determined and the distance from the source plotted against the original concentration. Figure 6 shows the relationship between chemotropism and the lowest and highest concentration between which chemotropism is observed, as a function of the distance from the donor block. It is interesting to note that in this experiment a concentration–response relationship is demonstrated for chemotropism which is very similar to the one obtained for chemotaxis (cf. Fig. 3). Clearly, this approach holds promise for the clarification of the mechanism of growth orientation.

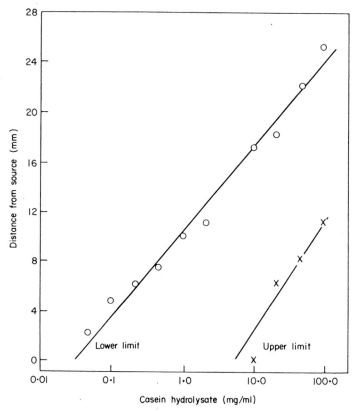

Fig. 6. Relationship between casein hydrolysate concentration and the distance from the source at which chemotropic growth can be observed in *Achlya* germ tubes. The furthest distance at which chemotropism is visible is represented by the plot of "lower limit" and the nearest point to the source of casein hydrolysate by the plot of "upper limit" (A. Musgrave, unpublished work).

2. Sirenin, a Chemotactic Hormone of the Fungus *Allomyces*

Members of the genus *Allomyces* (Order Chytridiales) are aquatic fungi, growing saprophytically on plant and animal material in nature. They can be easily cultivated in the laboratory on simple nutrients. The sub-genus *Euallomyces,* to which the best-investigated species belong, is characterized by a more or less regular alternation of a haploid, gameto-phytic, and a diploid, sporophytic phase, as is shown in Fig. 7. Both

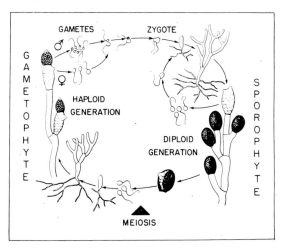

FIG. 7. Life history of *Allomyces macrogynus* (from Machlis *et al.*, 1966).

phases exhibit a branched mycelium, provided with rhizoids which fix the organism in the substrate. The gametophytic phase produces male and female gametangia which occur in pairs, the orange-coloured males on top of the colourless female gametangia, or vice versa, depending on the species. The gametangia release motile and naked uniflagellate gametes. The male gametes are attracted to the female gametes. When male and female cells meet, they fuse to form motile, biflagellate zygotes which ultimately settle down and develop into the asexual sporophytic phase. The sporophyte bears two types of sporangia. Some sporangia produce diploid zoospores which develop into new sporophytes, whereas others release haploid zoospores after meiosis, which grow into gametophytic plants.

TWO TYPES OF CHEMOTAXIS IN *ALLOMYCES*

The zygotes and both types of zoospores are attracted towards a source of amino acids. This type of chemotaxis has been described by Machlis (1969a). A mixture of leucine and lysine is primarily responsible for this attraction. These amino acids are active at concentrations higher than 10^{-5}M. The gametes, on the other hand, are not chemotactically responsive to amino acids. However, within no more than 5 min subsequent to fertilization the zygotes respond to a mixture of leucine and lysine (Machlis, 1969b). Apparently, gametic fusion potentiates rapidly a mechanism which makes the cells responsive towards a concentration gradient of these amino acids. In view of the knowledge accumulated about the mechanism of bacterial chemotaxis, this system seems very interesting for further study (see page 16 ff).

In contrast, male gametes are strongly attracted towards female gametes in a highly specific way. Female gametes have only slight motility and after dispersal are soon surrounded by males, which results in the formation of clumps of cells. Also, male gametes cluster in the vicinity of unopened female gametangia. A direct proof that the clustering is a response to a chemotactic agent was obtained by Machlis, who showed that male gametes are attracted to and remain in the vicinity of a dialysis membrane that has on its other side the supernatant in which female gametangia had released female gametes. The attractant was named "*sirenin*" (Machlis and Rawitscher-Kunkel, 1967).

All species of *Allomyces* are monoecious. The consequence is that one gametophytic mycelium produces a mixture of male and female gametes. Of course, this is a disadvantage if one wants to study the mechanism of interaction between both types of gametes. Therefore, it was of crucial importance that interspecific hybrids were developed which produced either largely male, or largely female gametes (Machlis, 1958). The work described in the following sections was done with hybrids of *Allomyces macrogynus* and *A. arbuscula*.

ISOLATION OF SIRENIN

The biological production of the sex attractant sirenin produced by female gametes of *Allomyces* hybrids was described by Machlis and co-workers in 1966. Ten litre batches of a solution containing only glucose and sodium acetate were inoculated with a predominantly female strain by daily adding fragmented mycelium grown for 56 h in 300 ml medium containing glucose and yeast extract. The cultures were incubated with vigorous aeration for 48 h. After that time the mycelia were covered with gametangia. These are formed only when growth is limited by a nutrient deficiency (in this case presumably nitrogen). The cultures were then diluted with an equal volume of water, which caused the release of gametes from the gametangia. After another 18 h the cultures were harvested. The mycelia were removed by filtration through cheesecloth and the resulting solution extracted with methylene chloride.

Since the chemotactic agent appeared to be produced in very low amounts (approximately 10^{-6} M), the accumulation of a sufficient amount of extract for elucidation of its structure necessitated a production scheme of semi-industrial proportions. Daily four 10 litre cultures were started and the resulting 80 litre medium processed until 2·5 g of sirenin-containing material had accumulated from a total of 10^4 litres culture medium (Machlis *et al.*, 1966).

BIOASSAY OF SIRENIN

A strong point in Machlis' investigations was his successful bioassay procedure, which allowed quantitative assessment of hormone concentrations. It was modified slightly through the years. The following description stems from Machlis (1973b). A steel tube (6 mm internal diameter, 9 cm long), the bottom sealed with uncoated cellophane, is mounted vertically in a Petri dish, so that the bottom is 3 mm up from the floor of the dish. The well above the membrane is filled with test solution until a positive meniscus is formed, and then covered with a glass coverslip. Into the Petri dish a suspension of male gametes is pipetted, covering the bottom of the steel tube. The test solution diffuses downward through the membrane resulting in a concentration gradient in the lower liquid containing the gametes. These react by swimming towards the membrane and adhere firmly to it. With a microscope the number of the gametes which are attached to the lower side of the membrane can then be counted. The number of gametes per unit surface after a specified time is a measure of the attractant concentration above the membrane. A strict timing of the assay is necessary, because the number of gametes attached to the membrane increases with time for about 90 min and then declines. Also the gamete density in the lower liquid is an important variable, the optimal value being 5×10^5 gametes per ml. For a proper response the gametes must be suspended in a well-buffered solution, containing Ca^{2+} ions and trace elements.

PURIFICATION AND PROPERTIES OF SIRENIN

Using this bioassay, sirenin was purified from the crude methylene chloride extract obtained by large-scale cultivation of strongly female *Allomyces* mycelia. A combination of chromatography on alumina columns and thin-layer chromatography on silica gel resulted in a highly purified preparation. In the older work an important additional step was esterification with 4-(4-nitrophenylazo)benzoyl chloride and repeated crystallizations, followed by hydrolysis of the derivative (Machlis *et al.*, 1966). The purified preparation was a colourless, optically active, viscous oil with a molecular formula of $C_{15}H_{24}O_4$, assigned from mass spectral data. Subsequent analysis of the product by nuclear magnetic resonance (n.m.r.) and infrared (i.r.) spectroscopy established the following structural formula for sirenin (Machlis *et al.*, 1968; Nutting *et al.*, 1968).

Structure of Sirenin

As suggested by Jaenicke (1972), the biosynthesis of this structure probably proceeds via *cis*-farnesyl pyrophosphate. Ring closure of this intermediate leads to a carene derivative which is oxidized to give the appropriate functional groups. Figure 8 summarizes the presumed biosynthetic pathway.

Racemic sirenin was obtained by total synthesis in several laboratories (Corey *et al.*, 1969; Grieco, 1969; Mori and Matsui, 1969; Plattner *et al.*, 1969; Bhalerao *et al.*, 1970; Corey and Achiwa, 1970). Plattner and

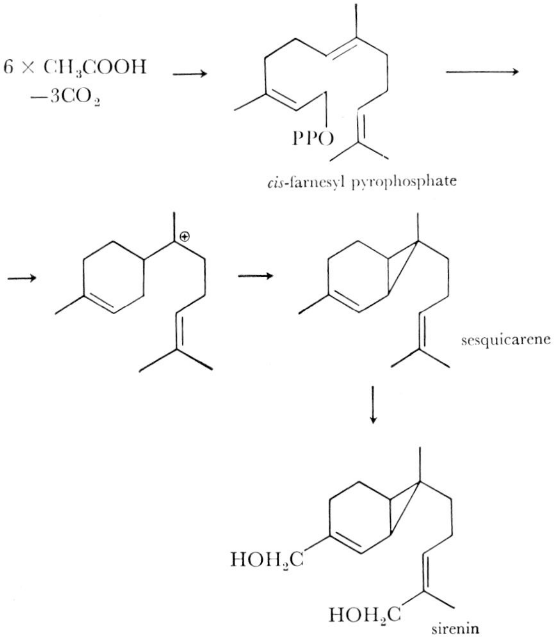

FIG. 8. Proposed biosynthetic pathway leading to sirenin in *Allomyces* (from Jaenicke, 1972).

Rappoport (1971) were able to obtain the separate enantiomers D- and L-sirenin. Only the L-enantiomer was biologically active.

TABLE I. Response of a range of concentrations of male gametes to various concentrations of sirenin (from Carlile and Machlis, 1965)

Male gametes/ml	Sirenin concentration (M)[a]							
	0	10^{-10}	10^{-9}	10^{-8}	10^{-7}	10^{-6}	10^{-5}	10^{-4}
10^6	4	35	159	365	∞	∞	∞	71
5×10^5	1	14	72	246	573	∞	∞	20
2×10^5	0	0	22	82	148	413	301	4
10^5	0	0	5	12	58	238	64	2
5×10^4	0	0	1	0	17	169	7	2

[a] The figures indicate the number of male gametes settled on 0·13 mm² of membrane after 1 h of chemotaxis. ∞ means too great to be counted accurately.

As in bacterial chemotaxis, there is an optimal concentration at which sirenin evokes the most vigorous response in Machlis' bioassay. Carlile and

Machlis (1965) estimated this concentration to be 10^{-6}M, as is illustrated in Table I. This table also shows that by using high concentrations of male gametes in the assay procedure, a chemotaxis is detectable with 10^{-10}M sirenin.

SPECIFICITY OF ACTION OF SIRENIN

As described above, sirenin was obtained from a highly female hybrid of *Allomyces arbuscula* and *A. macrogynus*. Male gametes obtained from a similar, but predominantly male hybrid respond vigorously towards this substance. The question arises if other species of *Allomyces* produce the same chemotactic substance, and also if gametes of other species do respond to sirenin. These questions have not been satisfactorily answered. Machlis (1968) only investigated the response of male gametes of the parent strains of *A. macrogynus* and *A. arbuscula*. Male gametangia of these monoecious species were collected from gametophytic mycelium by means of a micropipette. With the gametes so obtained, the bioassay procedure with sirenin was carried out. It appeared that male gametes of *A. macrogynus* reacted strongly towards this hormone, similar to those obtained from a hybrid strain. In contrast, the male gametes of *A. arbuscula* were at least ten times less responsive. According to Machlis, this poor response might be explained in two possible ways. One can assume that each species produces its own characteristic sirenin, with corresponding specificities in the response of the respective male gametes. This would imply that the hybrid progeny of *A. arbuscula* and *A. macrogynus* produces the same sirenin as *A. macrogynus,* and male gametes similar to those of the same species, with respect to response specificity. Alternatively, one can assume that both species produce the same sirenin, but that for some obscure secondary reason male gametes of *A. arbuscula* are less responsive. No data are available to support either of both possibilities. As Machlis points out, experiments of the type described are difficult to do. The major obstacle is the small number of male gametangia that can be collected from a mixture of males and females in one experiment, and the resulting low concentrations of gametes that are obtained for bioassay.

A second aspect of sirenin specificity was studied by testing the activities of synthetic analogues in comparison with the natural L-sirenin (Machlis, 1973a). It appeared that none of the following analogues showed any chemotactic activity in Machlis' bioassay. This suggests that both allylic alcohol functions as well as the sterical properties of the molecule are essential for biological activity. This is also apparent from a comparison of stereoisomers of synthetic sirenin and natural sirenin. The activities of these isomers are shown in Table II. D-Sirenin has virtually no activity at 5 nM compared to the L-enantiomorph. A mixture of both isomers shows a less than expected activity, suggesting that the mixture is not exactly racemic. Another explanation, that the D-isomer is a competitor of L-sirenin, and inhibits the action of the latter, was ruled out by demonstrating that addition of the D-isomer to a known amount of L-sirenin did not diminish its biological activity. The low activity of natural L-sirenin in comparison with synthetic L-sirenin could be explained by the presence of inactive or inhibitory impurities. It thus appears that synthetic L-sirenin is the material of choice to be used in future studies on the mechanism of action of this hormone.

TABLE II. Comparative chemotactic activity of various preparations of sirenin (from Machlis, 1973a)

Preparation	Concentration (nM)	Activity as percentage of 5 nM synthetic L-sirenin	
		Individual values	Average values
Synthetic L-sirenin	5·0	100	100
Synthetic L-sirenin	2·5	3, 6, 47, 67	50
Synthetic DL-sirenin	5·0	30, 34, 39, 42	36
Natural L-sirenin	5·0	23, 37, 45, 46, 66	43
Synthetic D-sirenin	5·0	4, 2, 4, 3	3
None		1, 2, 2	2

UPTAKE OF SIRENIN BY MALE GAMETES

When L-sirenin is incubated with a dense suspension of male gametes, it is rapidly consumed. A 5 nM concentration in a suspension of 5×10^5 gametes per litre is virtually reduced to zero within 20 min. Up to 100 nM the rate of uptake is constant and follows first-order kinetics. However, at higher (and quite unnatural) concentrations the uptake is slower and stops after 15 to 20 min. No sirenin can afterwards be detected in gametes which have absorbed the hormone, by extraction with organic solvents or assay of sonified gamete suspensions. This suggests that sirenin is metabolized by male gametes (Machlis, 1973a).

The question whether the uptake of sirenin is related to chemotactic action has not been answered. When male gametes have taken up sirenin from a 5 nM solution exhaustively, their ability to absorb more sirenin remains virtually unaltered, as is evident from the rate of uptake after addition of more sirenin. However, the responsiveness of these gametes toward a concentration gradient is severely diminished after a 20 min stay in a 5 nM solution. Only 50 min later the gametes respond normally again. This suggests that the ability to absorb sirenin is not strictly correlated with the response of male gametes toward a concentration gradient of this chemotactic hormone (Machlis, 1973a).

3. Sex Hormones in the Watermould *Achlya*

Achlya is one of the best-documented fungi with regard to sexual physiology. Its study was initiated by de Bary in 1881 and perpetuated almost
exclusively by Raper and Barksdale from 1939 till the mid-1960s. Their
work has been so successful that in more recent times *Achlya* has become
a popular object for developmental physiologists.

It is a representative of the Saprolegniaceae, aquatic fungi with hyphae
containing numerous nuclei. Crosswalls (septa) are practically only formed
to delimit sexual structures.

Most strains of *Achlya* reproduce sexually by self-conjugation, although
some behave as typical males or females in cross-conjugation. Generally,
however, there is no commitment to a particular sexual character, since
frequently a capacity to react either as male or female is exhibited depending on the sexual partner ("a sexual ambivalence in which maleness and
femaleness are determined in each mating by common consent of the
mated", Raper, 1959). I will return to this interesting feature in a later
section.

Let us first look at the sexual reactions between a typical male and
female strain, as can easily be done in a Petri dish, to get an accurate
picture of the events that take place.

DEVELOPMENTAL COURSE

The first visible response when a male and a female mycelium come near
to each other is the formation of branched antheridial hyphae on the male.
These branches are easily discernible from vegetative hyphae, being thin-
ner and less regularly shaped. They grow towards oogonial initials which
develop on the female thallus as spherical organs (Fig. 9). Antheridial
hyphae curve around the oogonial initials, and after a while cross walls

Fig. 9. Antheridial hyphae of *Achlya bisexualis* growing towards oogonial initials (×94).

are formed near the tips, the delimited cells being called *antheridia*. At about the same time a cross wall is formed at the base of each oogonial initial. Within the *oogonia* thus formed the protoplasm reorganizes to form 1–20 mononucleate eggs per oogonium. The antheridium penetrates the oogonial wall with mononucleate fertilization tubes, each of which fuses with one egg (Olive, 1953). It is believed that meiosis takes place in the sex organs and not in the fertilized egg, the vegetative mycelia being diploid (Barksdale, 1966).

One of the most striking features of the sexual morphogenesis is the sequential pattern, suggesting a successive secretion of stage-characteristic substances which control the morphological changes (Raper, 1952). Furthermore, sexual coordination is very similar in cross- as well as in self-conjugation, which, together with the occurrence of interspecific mating, indicates a common regulatory mechanism for at least the whole genus *Achlya*.

HORMONAL INTERACTIONS

The fact that a good deal of the sexual events in *Achlya* are based on hormonal secretions was demonstrated by a famous experiment designed by Raper (1940). He placed male and female mycelia in separate small cells; a stream of water was drawn sequentially through the cells via small connecting siphons. In the first cell a female mycelium was placed, in the second a male, in the third a female again, and in the fourth a male. The first response was observed in the second cell a few hours after the start of the experiment, where the male was seen to react by the formation of antheridial branches. Some hours later the male in the fourth cell was reacting vigorously in the same way. Comparatively long after the first male responses the female mycelium in the third cell reacted by producing oogonial initials which did not develop further. Finally, the antheridial hyphae on the male mycelium in cell 4 grew towards the tip of the siphon from cell 3. The antheridial hyphae in cell 2 did not grow towards the siphon from cell 1, nor did the female mycelium in cell 1 develop oogonia.

Raper concluded from these experiments that: (a) initiations of the entire sexual process depends on the secretion by the female of one or more substances which induce antheridial hyphae on the male mycelium (hormone A); (b) the production of oogonial initials on the female depends on a substance secreted by the male only after its activation by the female (hormone B); (c) the attraction of antheridial hyphae depends on a substance secreted by the female only after the production of oogonial initials (hormone C); the latter hormone could also function by inducing the delimitation of antheridia in branches which are in contact with oogonial initials; (d) a hormonal factor is produced by the male that stimulates the delimitation and further development of the oogonia (hormone D) (cf. Raper, 1952).

Some uncertainty about the existence of the hormone C was expressed by Barksdale (1963a). She obtained results, suggesting that hormone A functions not only to induce the production of antheridial hyphae, but also to orient the growth of these hyphae towards the oogonial initials and to bring about the formation of antheridia (septation). A chloroform extract of culture medium of *Achlya ambisexualis* T5, a typical female strain, was fractionated by means of countercurrent distribution. An active preparation thus obtained was adsorbed to polystyrene particles, which were sprinkled over the mycelium of *A. ambisexualis* E87 (a typical male strain) in three dishes. To the first dish free extract was added 2 h before the addition of the plastic particles, to the second dish it was added simultaneously with the polystyrene beads, and it was omitted from the third dish. The dishes were examined 4 h after the addition of the plastic particles. The antheridial branches produced ubiquitously in the two dishes containing additional free extract were "attracted" to the treated beads, whereas in the third dish branches were initiated only in the vicinity of each particle but were also attracted. No mention was made of a more distinct chemotropic reaction in dish 3 as compared with dishes 1 and 2.

When a square of cellophane was placed on the mycelium of E87 (male) and extract-treated particles were spread over it, many of the antheridial branches induced near the particles but on the other side of the membrane formed cross walls. Cross walls were also observed in antheridial hyphae suspended free in culture medium containing the extract. This contradicts an earlier suggestion of Raper, who conceived that for the delimitation of antheridia not only hormone C was necessary, but also the attachment of the antheridial branches to the oogonial wall or some other surface ("thigmotropic stimulus"; Raper, 1952).

The suggestion that hormone A also provides the chemotropic stimulus for the antheridial hyphae implies that in order to create a proper concentration gradient in the medium to attract antheridial branches, the oogonial initials produce more of this hormone than vegetative hyphae. The obvious experiment to test this would be to add hormone B to a female individual, thus inducing the formation of oogonial initials, and see if this would lead to the stimulation of hormone A production. An experiment of this type has not been described, presumably owing to the lack of hormone B. It is worth noting that in some matings described by Raper (1939) antheridial branches appear also to be attracted by the vegetative hyphae of the female.

When we consider Barksdale's expression of doubt over the existence of hormone C in conjuction with Raper's experimental results, it is clear that our conclusions should depend on the homogeneity of the hormone A preparations Barksdale used. If they consisted only of hormone A as the active factor, she was right in suggesting that this substance not only induces antheridial branches, but also orients their growth and promotes cross wall formation in the sexual organs. Final conclusions had to await further purification of hormone A. This was not an easy enterprise, since the substance, being extremely active, was only produced in minute quantities by vegetative mycelia. A good assay procedure had to be developed before further progress was possible.

BIOASSAY OF HORMONE A (ANTHERIDIOL)

The following description of a bioassay procedure is taken from Barksdale (1963b). Ten ml aliquots of medium inoculated with zoospores of the male *Achlya ambisexualis* E87 strain are aseptically pipetted into sterile plastic Petri dishes, and incubated at 25°C for 72 h. To each 1 ml sample to be assayed, 9 ml of distilled water is added. From this initial dilution, 1 : 3 and 1 : 10 dilutions are serially made. The three highest dilutions in each series are assayed. One ml of diluted sample is added to a dish of E87 and incubated for 2 h. If branching has occurred on 25% or more of the hyphae nearest to the surface, the dilution is recorded as active. A unit of hormone A is defined as the smallest amount in 1 ml to which strain E87 will respond within 2 h at 30°C by the production of antheridial initials (see Table III).

TABLE III. Effect of various concentrations of nutrient and antheridiol on the mean number of branches initiated per hypha in *Achlya ambisexualis* E87 (from Barksdale, 1969)

Dilution of E–G solution[a]	Mean number of branches			
	Antheridiol units per millilitre			
	1000	300	100	30
Undiluted	36·0	27·1	24·2	17·6
1:10	33·3	23·9	19·2	13·8
1:10²	34·1	21·3	13·2	10·7
1:10³	30·1	13·6	9·2	8·0
1:10⁴	22·2	8·1	6·4	6·2
1:10⁵	18·4	7·1	4·8	6·1
Salt solution[b]	8·1	5·3	4·6	4·5

[a] E–G solution contains hydrolysed lactalbumin, 400 mg; glucose, 2400 mg; calcium glycerophosphate, 80 mg; tris (hydroxymethylaminomethane), 100 mg; $MgSO_4 \cdot 7H_2O$, 125 mg; KCl, 150 mg; trace elements; distilled water, 1 litre.
[b] E–G solution minus hydrolysed lactalbumin and glucose.

The concentration of hormone A determines both the number of branches initiated and the time that elapses between the addition of hormone and the appearance of branches. The number of branches increases with increasing concentration of hormone until an upper limit is attained. The length of time before branches appear decreases with increasing concentration until a minimum of 40–45 min is reached (Barksdale, 1969). This assay, being very sensitive and having an operable dose–response relationship, is not very specific, however. Amino acid mixtures also induce branches in *Achlya* (Fischer and Werner, 1955).

ISOLATION AND CHARACTERIZATION OF HORMONE A (ANTHERIDIOL)

For large scale production, *Achlya bisexualis* T5 was cultivated in Fernbach flasks containing 360 ml of medium (McMorris and Barksdale, 1967). A weekly batch of 240 flasks gave 85 litres of culture liquid, containing about 3×16^6 units of hormone per litre. This was extracted with 13 litres of methylene chloride. After evaporation of the solvent a brown gum remained. This was purified by countercurrent distribution (later on by silica gel chromatography; Barksdale, 1969). Crystallization yielded about 10 mg of colourless crystals, melting at 250°–255°C (indicating a reasonable degree of purity), with a biological activity of $1 \cdot 5 \times 10^8$ units per mg. A concentration of 2×10^{-8} mg per ml was sufficient to induce branching, but 10^{-6} mg per ml was required to delimit antheridia. These figures indicate that the concentration per litre culture fluid was about 0·02 mg.

Once hormone A was obtained in crystalline form, it was renamed *antheridiol*. Its structure was elucidated by Arsenault *et al.* (1968) and Edwards *et al.* (1969) by spectroscopy and total synthesis. The configuration at C-22 and C-23 was proved to be 22*S*, 23*R* by Edwards *et al.* (1972).

Structure of antheridiol.

Four stereoisomers and several intermediates, differing from antheridiol mainly in the structure of the side chain and functional groups were tested for the ability to evoke a response from *Achlya ambisexualis* E87, the most sensitive strain known (cf. Barksdale *et al.*, 1974 for references). It appeared that the stereochemistry at C-22 and C-23 in antheridiol is most important. Table IV shows the minimal concentration of the four

TABLE IV. Minimal concentration of stereoisomers of antheridiol inducing branching in *Achlya ambisexualis* E87 (from Barksdale *et al.*, 1974)

Isomers	Configuration of side chain of antheridiol[a]	Minimal concentration inducing branching (ng/ml)
Antheridiol (22*S*, 23*R*)	OH / R / O / O	0·006
Erythro isomer (22*R*, 23*S*)	OH / R / O / O	20
Threo isomer (22*S*, 23*S*)	OH / R / O / O	20
Threo isomer (22*R*, 23*R*)	OH / R / O / O	400

[a] R = Complete four-ring skeleton of antheridiol.

possible stereoisomers inducing branching. The 22*R*, 23*S* isomer is less than 1/1000 as active as antheridiol, whereas the others are even less active. The authors do not preclude the possibility that contamination of the three isomers with antheridiol is responsible for this low activity. The carbonyl group at C-7 in the B ring appears to be relatively unimportant, since antheridiol is only twenty times as active as its 7-deoxy analogue. Also, a free hydroxyl at C-3 in the A ring is not essential because acetylation does not decrease the biological activity. A large number of other derivatives and mammalian hormones proved to be inactive (Barksdale *et al.,* 1974).

ACTION OF ANTHERIDIOL

From the foregoing discussion four functions can be ascribed to antheridiol: (1) induction of antheridial hyphae; (2) chemotropic stimulation of these hyphae; (3) stimulation of male mycelium to produce hormone B; (4) induction of septum formation in antheridial hyphae. These and some other aspects of hormonal action in *Achlya* will now be discussed separately.

Induction of Antheridial Hyphae

The problem of how a substance of well-established structure evokes a specific response in a mycelium that is sensitive to it, is of considerable interest from a general developmental point of view. Very little work, however, has contributed to an answer. First of all it should be noted that antheridiol is not the only substance to induce branching. Fischer and Werner (1955) have demonstrated that a mixture of amino acids stimulates branching as well as chemotropically orients hyphal growth in *Achlya,* though only under poor nutritional conditions. They suggest that the attractive force of oogonial initials is due to amino acids, secreted during egg formation. They mention in their support that zoospores, which are chemotactically attracted to amino acids, are actively attracted to oogonia and become sessile on their surface, and that often antheridial branches make secondary branches in the neighbourhood of an amino acid source. This raises the question of whether antheridiol serves to induce specialized sexual structures or merely stimulates branching, in the same way that amino acid mixtures do. Barksdale (1970) emphasizes that antheridial branches are really different in form and function from branches evoked by amino acids. Antheridial branches, in contrast to vegetative branches, are slender and have a tendency to be spiralled (cf. Fig. 9). Moreover, it is only from branches induced by antheridiol that antheridia are delimited. She also found, however, that branching is a function not only of antheridiol concentration, but also of the amount of nutrient available. Omission of nitrogen and carbon sources dramatically reduces the number of branches and prolongs the time before branches appear, as is shown in Table III. Nevertheless, branching seems to be in the first place dependent on antheridiol concentration; the availability of nutrient determines whether a branch initiated by antheridiol will develop into a real antheridial branch from which an antheridium can originate, or into a vegetative branch. Conceivably the primary action of antheridiol is to initiate the formation of specialized branches; this initiation is stimulated by nutrient materials. At a high nutrient level, however, dedifferentiation could take place.

The concept that antheridiol induces sexual differentiation is strengthened by the results of Silver and Horgen (1974). They have established that antheridiol markedly stimulates [³H] adenosine incorporation into a fraction of RNA, presumably mRNA. After an initial lag of 2–3 h, a three to fourfold increase in the specific activity is observed. The enhancement of incorporation 3 h after addition of antheridiol corresponds to the observation of the appearance of antheridial branches. Addition of actinomycin D causes a rapid decline in the specific activity of a polyadenylate-rich fraction within 1 h, suggesting that the lifetime of this RNA fraction is very short. Also, protein synthesis is stimulated by antheridiol after a 3 h lag period, implying that this compound stimulates the synthesis of mRNA required for sex organ differentiation in *Achlya*. The same conclusion was drawn by Horowitz and Russell (1974). They found that actinomycin D effectively inhibited induction as well as elongation of antheridial branches, whereas the growth of vegetative hyphae remained unaltered.

It seems logical to assume that the formation of an antheridial branch is preceded by a localized hydrolysis of the rigid cell wall. This was convincingly demonstrated in another system, namely budding yeast (Matile *et al.*, 1971) where an increase of cell wall-hydrolysing glucanase during early stages of budding was found. This hydrolase was concentrated in a microsomal fraction that was thought to contain the vesicles centering near the point of bud formation (Moor, 1967). Also, in actively growing hyphal tips, vesicles are thought to contain cell wall-hydrolysing enzymes besides precursors and synthesizing enzymes (Bartnicki-Garcia, 1973). Thomas and Mullins (1969) have found that branching in *Achlya* is accompanied by an increase in cellulase. Antheridiol stimulates cellulase secretion in male mycelia, where it induces antheridial branches, but not in female mycelia. This enzyme is thought to affect the cell wall, which contains approximately 15% cellulose. However, an overall increase of cellulase does not account for the fact that branch formation is a very localized event. Mullins and Ellis (1974) have tried to meet this problem by demonstrating that aggregates of vesicles are found at the point of branch formation. As a corollary, Nolan and Bal (1974) have suggested that such vesicles do indeed contain cellulase. They incubated male *Achlya* mycelium, which was stimulated with antheridiol and subsequently fixed with a glutaraldehyde–formaldehyde mixture, with carboxymethyl-cellulose and demonstrated cytochemically that the reaction product of carboxymethylcellulose hydrolysis was deposited within vesicles and dictyosomes (the presumed origin of these vesicles). In summary, these studies imply that branch formation in *Achlya* proceeds in a fashion analogous to bud formation in yeast, cellulase being formed in the dictyosomes and transported to restricted, perhaps predetermined, areas of the cell surface, where it hydrolyses the cell wall as a starting point for branch formation. Unfortunately, though, we still do not know what makes the branch an antheridial branch, as opposed to a vegetative one.

Chemotropism

It is believed that antheridial branches grow chemotropically towards a source of antheridiol in the form of an oogonial initial or a plastic bead to which antheridiol is adsorbed, but proof of this requires the demonstration that repositioning of the source is immediately followed by a reorientation of pre-existing hyphal tips to distinguish it from growth stimulation (see Chapter 1). Such proof is lacking as yet. As a cautionary note it should be recalled that the antheridial branches may grow to the oogonial initials up a gradient of amino acids, as suggested by Fischer and Werner (1955).

Stimulation of Hormone B Production

As described on page 37, Raper (1940) observed that in a cross-conjugation the female mycelium is much slower in producing sex organs than the male. He suggested as an explanation that hormone B, responsible for female induction, is only produced by a male which is stimulated by antheridiol. Barksdale and Lasure (1973), trying to verify this, have found that *Achlya ambisexualis* E87 secretes hormone B only when stimulated by the addition of exogenous antheridiol. At the same time they indicate that this may not be a general feature of hormone B-producing strains, several of them being insensitive towards antheridiol.

Delimitation of Antheridia

There is disagreement in the literature as to the factors responsible for antheridium delimitation. Whereas Barksdale (1963a) states that formation of antheridia can be observed even when the antheridial branches treated with antheridiol are suspended free in the medium, we cannot ignore Raper's observation that antheridial hyphae, wherever they come into contact with a surface, produce at their tips characteristically branched swellings which become delimited as antheridia. Antheridium formation and differentiation are thus shown to require both contact with a definite surface and specific secretion(s) of ripened oogonial initials (Raper, 1952). More research is evidently needed to establish the conceived role of surface contact.

Degeneration

An interesting action of antheridiol is the degeneration of sexual structures in self-conjugating strains. Antheridiol not only strongly inhibits the formation of oogonial initials, but also causes a degeneration of oogonial initials when added to a mycelium having both male and female structures. When relatively high concentrations of antheridiol are added, oogonial initials at all stages of their development prior to septation nearly always dedifferentiate (Raper, 1950b). This may take place by a flow of the protoplasmic contents of the initial back into the vegetative hypha, or by the outgrowth of antheridial hyphae from the oogonial initial itself. Parenthetically, this action has also been demonstrated for high concentrations of amino acids (Fischer and Werner, 1955). It seems puzzling that oogonial initials which attract male sexual structures by secreting a certain substance, can be degenerated by it.

Uptake of Antheridiol by Male Strains

Antheridiol seems to be actively consumed by strains with a male char-

TABLE V. The rate of uptake of antheridiol by five strains of *Achlya* (from Barksdale, 1963b)[a]

Strain number	Initial concentration of antheridiol (units/ml)		
	10^5	3×10^4	10^4
E15 (strongly male)	1600	450	180
E22 (strongly male)	1600	400	180
E87 (strongly male)	660	200	100
9 (weakly male)	600	130	65
10 (weakly male)	330	90	30

[a] Rate of uptake expressed as units of antheridiol taken up per mg dry weight per min. The values given are the mean of those obtained in three experiments.

acter, and also by typical self-conjugating strains (Barksdale, 1963b). This finding was borne out by the discovery that the concentration of antheridiol in a female culture (like *Achlya ambisexualis* 734) was 1000 times greater than in a mixed culture 734 × E87). Table V shows some data indicating that the rate of uptake is not only dependent on the concentration of antheridiol present, but also on the strain that is used. It appears that the capability to produce antheridial initials approximately parallels the uptake and also the amount of antheridiol that is removed from the medium: strains E22, E15 and E87 that rapidly produce antheridial initials and can be classified as strong males take up antheridiol more readily than strains 9 and 10, which are considered as weak males because they are capable of producing oogonial initials in a sexual reaction with a strongly male strain. Typical female strains do not seem to take up antheridiol at all. The corollary is that antheridiol is only accumulated in the culture medium by strains with an outright female character (Barksdale, 1970).

Some doubt has been cast on these results, however, since preliminary studies have demonstrated that ^{3}H-labelled antheridiol is bound or taken up by male as well as female mycelia (Barksdale and Lasure, 1973). An explanation must await further reports on this matter.

The fate of antheridiol that is taken up by the male is unknown. Nevertheless, this problem is of special interest, in the first place because more information could tell us something about the primary action of antheridiol, and in the second place because the possibility comes readily to mind that antheridiol might serve as a precursor of hormone B, the oogonia-inducing substance secreted by the male in response to antheridiol.

OTHER HORMONES IN *ACHLYA*

Recently, two crystalline compounds have been isolated by McMorris and coworkers (1975) from culture liquids of *Achlya heterosexualis* which possess hormone B activity. The following structures have been proposed for these compounds, which have been named *oogoniol-1* and *oogoniol-2*. Oogoniol-1 and -2 showed similar activity in biological tests for hormone B. The lowest concentration at which activity could be observed was 620 and 460 ng per ml, respectively.

If it is true that antheridiol orients the growth of antheridial branches towards the oogonial initials, and also governs the delimitation of

Structure of oogoniol-1 ($R = (CH_3)_2CHC = 0$) and oogoniol-2
($R = CH_3CH_2C = 0$)

antheridia, it then becomes unnecessary to postulate the existence of a hormone C.

About the existence of hormone D, which was believed to play a role in oogonial delimitation, no direct evidence has been produced since it was postulated. Its existence was based on the fact that it takes 2–4 h of contact between antheridial and oogonial initials before the latter are delimited. The constancy of this time lapse is significant, and suggests that some substance is produced by the antheridium, after its delimitation, which is necessary before the oogonial initial is capable of becoming transformed into an oogonium (Raper, 1939).

Some more hormonal substances were conceived by Raper, designated together as the hormone A complex (Raper, 1950a). We must pay some

TABLE VI. Effect of male culture filtrate on the reaction of male mycelia to antheridiol (from Raper, 1942)

Antheridiol (units/ml)[a]	Control[b]	Male filtrate added[b]
0	0	0
1	0·6	11·8
2	1·7	12·6
3	2·7	13·1
4	3·0	12·4
5	3·6	10·4
6	4·2	12·5

[a] Units according to Raper are equivalent to 2–4 Barksdale units (Barksdale, 1963b).

[b] Reaction intensity is expressed as average number of antheridial branches.

attention to them because a reinterpretation of this older work might be possible in the light of more recent results.

Raper suggested that vegetative male mycelia secrete a factor, called hormone A[1] which influences the response of antheridiol. This is illustrated in Table VI, which indicates a strong positive influence of male filtrate upon antheridiol action. The properties of the supplementing factor (dialysable, neutral, non-migratory in an electric field, stable to heat, acid, oxidizing reagents, destroyed by prolonged alkaline hydrolysis) do not exclude the possibility that it actually consisted of a mixture of hydrolysis products secreted by the male mycelium. Nothing is present in Raper's work to indicate that female culture filtrates were also examined so there is no reason to presume that the factor is exclusively produced by male mycelium. Conceivably this A[1] factor consisted of an amino acid mixture, as suggested by Barksdale (1970) (see Table III).

A second factor postulated by Raper was hormone A[2], specifically produced by female mycelium, and, like antheridiol itself, capable of inducing antheridial branches in male mycelia. This factor was distinguished from antheridiol by its acetone insolubility. Raper's procedure was as follows: female culture filtrate was taken up in diatomaceous earth and the water evaporated *in vacuo*. The residue was then extracted with acetone (removing antheridiol) and subsequently with water. The water contained the A[2] factor. Additions of A and A[2] together to a male culture resulted in a more than additive response. This result led Raper to believe that A[2] was a specific entity. In retrospect, it is attractive to suppose (as Raper himself partially suggested) that the aqueous extract of the diatomaceous earth contained residual antheridiol and amino acids. The presence of the latter would provide an explanation of the more-than-additive effect when antheridiol and A[2] were added together. This interpretation (which needs experimental confirmation) would obviate the need for postulating on Hormone A[2].

The third factor in Raper's hormone A complex, hormone A[3], inhibited antheridia formation in male plants. It was extracted from male cultures by acetone according to the method described above. As with hormone A[1], no mention was made of extraction from female filtrates so that its sex specificity remains in doubt. This and its general inhibitory action leaves the possibility that it is a general growth inhibitor produced by male and female mycelia alike. Growth inhibitors appear to be quite common in fungi (Robinson, 1969).

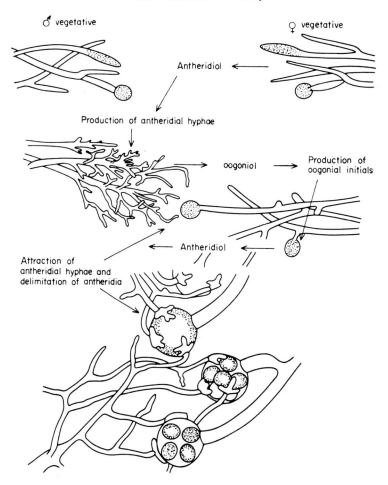

FIG. 10. Proposed role of antheridiol and oogoniol in sexual morphogenesis of *Achlya* (from Raper, 1955, adapted).

The hormonal regulatory system in *Achlya,* including the simplifications suggested above, is presented in Fig. 10.

GRADUATED SEXUAL RESPONSES IN *ACHLYA*

Most if not all species of the (sub)genus *Achlya* are monoecious, which means that oogonia and antheridia formed on one thallus can usually conjugate with each other. This does not exclude the possibility of cross-conjugation with other thalli. However, the tendency to produce antheridial and/or oogonial initials differs from one strain to another, and so does the ability for self-conjugation. For instance, strains of *Achlya bisexualis* and *A. ambisexualis* seldom self-conjugate, but within species they readily conjugate with one another. They can be arranged into grades between the two extremes "male" and "female". The frequency of self-conjugation is lowest at either end of the scale (Barksdale, 1965).

The expression of a male or female tendency is not constant. It is dependent on the mating partner, and also on the environmental conditions (Barksdale, 1960). Typical male strains of *A. ambisexualis* c11 and c5 behave as females when combined with e87; strain c5, a typical female, behaves as a male towards strain 734. Some oogonia are delimited on e247 when combined with 734 at 15° and 20°C but not at 25° and 30°C. Sometimes a redifferentiation of sexual organs is observed, for instance antheridial hyphae arising from oogonial initials, or an oogonium formed at the tip of an antheridial hypha (Raper, 1950b).

Reciprocal induction of sexual organs can occur between strains that tend to self-conjugate and strains that cross-conjugate. This suggests that also in self-conjugation the hormonal interactions take place which have been envisaged for conjugations between male and female thalli (Raper, 1950b). Indeed, Barksdale and Lasure (1973) have demonstrated that self-conjugating strains of *Achlya bisexualis* and *A. ambisexualis* secrete and respond to both antheridiol and hormone B.

It is worthwhile to pay some attention to *Achlya heterosexualis* with

TABLE VII. Sexual behaviour of seven strains of *Achlya heterosexualis* (from Barksdale, 1965, modified)

Strain	Minimal antheridiol concentration causing branching (units/ml)	Sexual reaction[a]						
		B14	B11	B1	B4	B15	B9	B13
B14	1·0		—	—	+	+	+	+
B11	10			—	+	+	+	+
B1	10				+	+	+	+
B4	10					+	+	+
B15	10						+	+
B9	5000							+
B13	5000							

[a] —, No reaction; +, strains in vertical row react as males with respect to strains in horizontal row.

emphasis on the hormonal aspects of the sexual reactions (Barksdale, 1965). Strains of these species are capable of both self-conjugation and cross-conjugation. In self-conjugation, antheridial branches arise from the same hypha as the oogonia, or from hyphae nearby. In contrast to intraspecific matings within *A. bisexualis* and *A. ambisexualis*, where antheridial branches are induced on one partner, followed by oogonial initials on the other, in *A. heterosexualis* only the antheridial branches are cross-induced. Oogonia are only initiated during self-conjugation (which occurs simultaneously with cross-conjugation). One can imagine that this sexual pattern is caused by either a very low productivity of hormone B or a low sensitivity to it. Also, strains of *A. heterosexualis* can be arranged in a series in order of decreasing male character, as shown in Table VII. Strains B14, B11 and B1 have a male character, producing antheridial branches in matings with the other strains; B4, B15, B9 and B13 readily show self-conjugation. Typical female strains appear to be absent. The male character of these strains is more or less reflected in their sensitivity towards antheridiol. Increasing femaleness goes with decreasing sensitivity. There does not seem to exist such a corollary with the productivity of antheridiol, which is very low in all strains of *A. heterosexualis*.

4. Sex Hormones in *Mucor mucedo* and Related Fungi

Sexual reproduction in the Mucorales was discovered by Blakeslee in 1904. He concluded from a survey of a great number of species that these fungi can be divided into two main groups, which he termed *heterothallic* and *homothallic,* respectively. In each species of the first group two strains or races could be distinguished, which, when grown apart, produce only sporangia (the non-sexual way of reproduction), but which conjugate and form sexual spores (*zygospores*) when their mycelia meet. Species of the second group form zygospores under suitable circumstances when grown from a single spore, without obligatory interaction with another mycelium.

The terms homothallic and heterothallic have been applied to other fungi since Blakeslee introduced them, but for reasons mentioned earlier (see Chapter 1) the terms *dioecious* and *monoecious* will be used here to denote these sexual systems.

Blakeslee was convinced that both strains of dioecious species were different sexes, but owing to the lack of morphological differences he designated them as *plus* and *minus* instead of male and female. He termed as *plus* that strain of a particular species which had the most luxuriant mycelial growth, and he was able to extend this nomenclature to other, related, fungi since frequently incomplete matings appeared to occur between different species. Conjugations were never observed between strains of the same mating type, and a change of mating type never occurred, indicating that this sexual characteristic has a firm genetic basis. This was confirmed by later workers (e.g. Kniep, 1929).

In a later publication Blakeslee and coworkers (1927) reported after a study of 2000 races included in twelve genera that never more than two mating types were found in dioecious species. Between 10 000 and 20 000 intraspecific combinations were made without getting evidence for sex intergrades. The same result emerged from a study of interspecific reactions (Blakeslee and Cartledge, 1927). The authors concluded: "The strict sexual dimorphism in the forms studied, as well as their morphological simplicity of sex differentiation, renders the group peculiarly adapted for use in an investigation of the fundamental differences between heterothallic sexes. The fact that in dioecious species all the races of one sex are theoretically capable of giving an imperfect reaction with all the sexually opposite races of every other species, while both sexes have been found to give imperfect sexual reactions with homothallic species, has led to the belief that there must be something fundamental common to all the *plus* races, for example, responsible for these reactions."

DEVELOPMENTAL COURSE

In Blakeslee's first study the species *Mucor mucedo* has furnished most of the information about the conjugation process because of its simplicity of sexual development *above* the substrate. Also in later work it has served as a model system, assumed to be representative of the whole order. The following description concerns this organism. When grown separately, there is no difference in the growth pattern between both mating types. They form aerial hyphae on a solid substrate which are slender and are readily distinguished from the stout hyphae destined to become sporangiophores. Where the mycelia of the opposite strains grow into close proximity, special more or less erect hyphae are produced which are intermediate in size between vegetative hyphae and sporangiophores. These hyphae are called *zygophores*. Zygophores of opposite mating type grow (or bend) towards each other by a process of mutual attraction. Their contact is seldom tip-to-tip. In some cases the point of contact is slightly back of the tips, or one zygophore is laterally met by the apex of the other. Sometimes both zygophores meet laterally. At the point of contact a swollen outgrowth called the *progametangium* rapidly develops on either zygophore. *Plus* and *minus* progametangia adhere firmly to each other, and, while enlarging, push apart the zygophoric hyphae from which they originated. When they have attained a considerable size, septa are formed which cut off more or less equal terminal cells, the *gametangia*. By dissolution of the wall at the site of contact both gametangia fuse and give

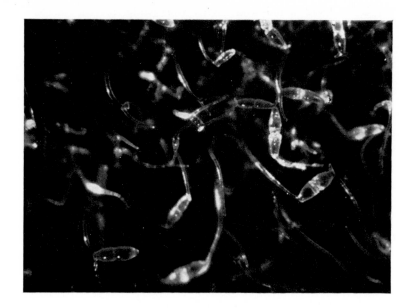

Fig. 11. Copulating zygophores of *Mucor mucedo* ($\times 100$).

rise to the zygote which rounds off and forms a dark and denticulated wall. The mature fusion cell is called the *zygospore* (cf. also Gooday, 1973).

This description, mainly derived from Blakeslee (1904), and depicted in Fig. 11, is applicable in its major characteristics to several other dioecious species. Often, however, as in *Mucor hiemalis* and *Absidia glauca,* no clearly discernible zygophores are formed. *Blakeslea trispora* and *Phycomyces blakesleeanus,* among others, differ in so far as primary contact takes place under the surface of the substrate. In the first species the zygospores are formed at the site of first contact, while in the latter the zygophores grow upward after contact and produce the zygospores above the substrate.

The sexual process in the dioecious *Mucor mucedo* is also similar to that in monoecious species. Zygophores, sometimes easily distinguished from non-sexual structures, are produced and form zygospores after fusion above the substrate. A couple of zygophores may originate from the same branch, as in *Zygorhynchus exponens,* or from separate branches, as in *M. genevensis.* In *Z. moelleri,* a side branch conjugates with the hypha it originated from. Also in these monoecious species, an attraction between conjugating filaments seems obvious at prolonged observation (see Fig. 16).

The fine structure and development of the zygospore of the monoecious *Rhizopus sexualis* has been investigated by Hawker and coworkers (1968) and Hawker and Beckett (1971).

As mentioned above, zygospores are formed spontaneously in monoecious species. To this it must be added that sometimes, under environmental conditions inhibiting zygospore formation, the close proximity of another, but dioecious, species stimulates conjugation. For example, the monoecious *Cunninghamella echinulata* will readily participate in a sexual reaction with dioecious species of *Mucor* at temperatures below 20°C, but will not itself form zygospores at low temperatures (Blakeslee, 1920). An agamic strain of *Zygorhynchus moelleri*, unable to produce zygospores, does so in the proximity of *Mucor mucedo* mycelium (Schipper, 1971).

HORMONAL INTERACTIONS

Investigations in the physiological background of sex in Mucorales were given a solid basis by the work of Burgeff. In an extensive and classic publication he demonstrated in 1924 that in *Mucor mucedo* and many other species sexual differentiation is governed by low-molecular-weight substances which diffuse through the substrate and the atmosphere and not only have an inductive action on zygophore formation but also direct their growth, leading them to sexual contact. It is worthwhile to discuss some of his findings. When two opposite strains of *M. mucedo* grow towards each other on a solid substrate, a declining growth rate in the most advanced hyphae is observed. When there is still a zone of uncovered agar between the mycelia, zygophores are produced on both of them. This indicates a mutual influence other than by cell-to-cell contact. The following experiment substantiates this conclusion. A mycelium of one mating type was covered with a collodion membrane on top of which was placed a piece of agar with mycelium of the opposite mating type, in inverted position. The latter was allowed to grow over the membrane. After 48–72 h one could observe numerous zygophores on the upper mycelium which for the greater part were curved downward to the membrane. Apparently the lower mycelium was able to induce the development of zygophores in the upper mycelium through the membrane. Emanations from the lower mycelium had forced the zygophores to orient towards the membrane. The described experiment, which is illustrated in Fig. 12, is equally successful if plastic or cellulose dialysis membranes are used, irrespective whether the *plus* or the *minus* mating type is underneath.

Similar observations, implying a participation of diffusible sexual hormones, were made by Burgeff in other dioecious species, like *Mucor hiemalis* and *Phycomyces blakesleeanus*. Especially *M. hiemalis* is worth mentioning because this species provided the best evidence for the exist-

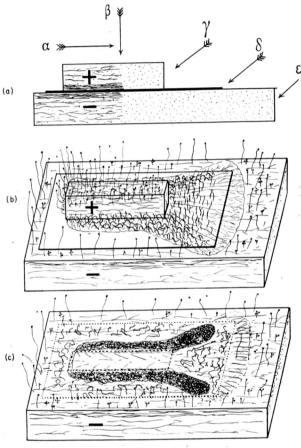

Fig. 12. Schematic representation of a membrane experiment with *Mucor mucedo*. (a) Position at the start of the experiment: α, direction of growth; β, front zone of the mycelia; γ, agar block; δ, collodion membrane; ε, agar substrate in a Petri dish. (b) Development of the mycelia after approx. 48 h. Zygophores on the upper mycelium induced by the lower mycelium. Note that many zygophores bend downwards to the membrane. (c) Lower mycelium after removal of the membrane and upper mycelium with swollen, strongly coloured hyphae in the region where zygophores have been present at the upper side of the membrane. Outside this zone many distorted sporangiophores, grown under the membrane, are shown (from Burgeff, 1924).

ence of volatile zygophore-inducing substances. It produced zygophores even when diffusion-contact with the partner through the substrate was inhibited, and mutual influence was only possible through the air. The occurrence of volatile, zygophore-inducing substances in dioecious and monoecious species wat later confirmed by Hepden and Hawker (1961) and Mesland *et al.* (1974).

All fungi mentioned so far also reacted vigorously in interspecific combinations, suggesting that they were all sensitive towards and also produced the same hormones. *Absidia glauca* and *Rhizopus stolonifer* seemed aberrant in this respect in producing zygophores only after accidental contact. However, these species appeared to induce numerous zygophores in *Mucor mucedo* when approaching this fungus. *Rhizopus* stolones (long hyphae, connecting two groups of rhizoids), when growing near or over *M. mucedo* mycelium, induced rows of zygophores on the latter which tried to make contact by bending towards them. *A. glauca,* normally not displaying oriented growth, sought actively contact with *M. mucedo* zygophores of the opposite mating type.

Burgeff concluded from this survey that all fungi investigated produce identical or very similar mating-type specific hormones which induce the formation of zygophores and cause directed bending of these structures. The regular conjuncture of zygophore induction and oriented growth made Burgeff (1924) suggest that both phenomena are caused by the same factors.

BIOASSAY OF ZYGOPHORE-INDUCING HORMONES

Banbury (1954) and Plempel (1963) were able to show that *plus* and *minus* mycelia of *Mucor mucedo,* grown together in liquid culture, accumulate substances in the culture medium which induce zygophores in both strains of the same fungus. The following assay procedure was developed to test the activity of these substances (Plempel, 1963). *Mucor mucedo-plus* or *-minus* is inoculated excentrically in Petri dishes containing nutrient agar. After 3 days, when about half of the plate is covered, test solutions are pipetted in wells just in front of the advancing tips. The plates are kept overnight in the dark, and the zygophores developed are observed and counted the next morning, using a stereomicroscope. With some precautions zygophore counts yield quantitative data about hormone concentrations, but this has not been practised extensively, primarily

c

because of the variability of the strains used, and the lack of proper standard solutions (van den Ende and Stegwee, 1971).

ISOLATION AND PURIFICATION OF ZYGOPHORE-INDUCING HORMONES

Plempel (1963) demonstrated that the active factors could be isolated by extraction with organic solvents from acidified culture medium of *plus* and *minus* mycelia, grown together. He also determined their chemical properties to a large extent, and could confirm the conclusion drawn by Burgeff (1924) that apparently identical factors were produced by different dioecious species.

The structural characterization of the hormonal factors was accomplished by Caglioti *et al.* (1967) and Cainelli *et al.* (1967). These investigators had become interested in the sexual system of Mucorales because of the observation that a chloroform extract of submerged mated cultures of *Blakeslea trispora* contained acidic substances which, when added to cultures of the single strains of this fungus, considerably increased the yield of carotenoids (see Ciegler, 1965, for review). These substances, named *trisporic acids,* were shown by van den Ende in 1968 to be identical with the zygophore-inducing substances. The structural formulae of the trisporic acids B and C are the following :

Trisporic acid B   Trisporic acid C

The trisporic acids B and C each comprise two geometrical isomers (at C-9) which can be separated as the methyl esters (Bu'Lock *et al.*, 1972). Trisporic acid A, which probably lacks the functional group in the side chain, occurs in very small quantities in culture media of *B. trispora* (Bu'Lock *et al.*, 1972). Because it is supposed to be a transformation product of trisporic acid C, with low biological activity, it is usually ignored.

The stereochemical characteristics were established by Reschke (1969), Austin *et al.* (1970) and Bu'Lock *et al.* (1970). Racemic 7-*trans,*9-*trans-*

trisporic acid B methyl ester was synthesized by Edwards *et al.* (1971). The compound showed after hydrolysis an activity equal to natural trisporic acid B from *Blakeslea trispora,* eliciting zygophore formation in both mating types of *Mucor mucedo.* The natural product, however, is optically active (van den Ende and Stegwee, 1971). Isoe *et al.* (1971) and White and Sung (1974) also described the total synthesis of trisporic acids.

It was demonstrated by van den Ende (1968) and Austin *et al.* (1969a) that trisporic acids from *Blakeslea trispora* and *Mucor mucedo* are indistinguishable, the only difference being the yield: *M. mucedo* produces about 1–10 μg/g dry weight, whereas the production rate in *B. trispora* may be as high as 1 mg/g. The latter is therefore the organism of choice for the production of trisporic acids.

The trisporic acids B and C are oily substances with a characteristic odour. They are quite susceptible to light, oxygen and extreme pHs (van den Ende, 1967). They exhibit a typical ultraviolet (u.v.) spectrum, with maximal absorption at 325 nm. The long wavelength of this absorption maximum is attributed to the presence of a trienone system, strongly perturbed by the non-conjugated substituents at C-1 (Bu'Lock *et al., 1972*).

BIOLOGICAL PROPERTIES OF THE TRISPORIC ACIDS

The most remarkable property of the trisporic acids is their ability to elicit zygophore formation in *both* mating types of *Mucor mucedo* (van den Ende, 1968; Gooday, 1968). This lack of mating-type specificity is not paralleled by other plant sex hormones studied thus far. In any other system a sex hormone, produced by one mating type, is exerting its action on the opposite mating type. Table VIII shows the results of comparative bioassays on *M. mucedo plus* and *minus.* It is clear that both mating types

TABLE VIII. Comparative bioassays on *plus* and *minus* strains of *Mucor mucedo* (from Bu'Lock *et al.,* 1972)

Compounds	Zygophores/μg	
	Plus	*Minus*
9-*cis*-Trisporic acid C	150	550
9-*trans*-Trisporic acid C	150	420
9-*cis*-Trisporic acid B	260	910
9-*trans*-Trisporic acid B	160	450
Trisporic acid A	50	120

are sensitive to all of three compounds, 9-*cis*-trisporic acid B being the most active one (Bu'Lock *et al.*, 1972). The difference in response between the *plus* and the *minus* strains is probably not biologically significant. Comparable results were obtained by other investigators (Gooday, 1968; van den Ende, 1968).

The trisporic acids are not only active in *Mucor mucedo,* but also in other, related, fungi. Thus zygophores are induced in *M. hiemalis,* which is not very surprising, considering the extensive illegitimate conjugative behaviour between *M. mucedo* and *M. hiemalis.* More notable is the stimulating effect trisporic acids have in monoecious organisms, like *Syzigites megalocarpus* (Werkman and van den Ende, 1974) and *M. genevensis,* which show a strongly increased zygosphore production in response. This suggests that the trisporic acids control sexual differentiation in a number of monoecious and dioecious Mucorales. Their role is not exclusive however, because several other metabolites produced by mated *M. mucedo* induce the formation of zygophores in a sex-specific fashion, as will be described in the next section.

The second effect which characterizes the trisporic acids is their stimulative action on carotenoid production in many mucoraceous fungi. This action has been investigated most intensely in *Blakeslea trispora.* In the older literature (reviewed by Hesseltine, 1961) several authors noted a stimulation of carotenoid synthesis during sexual activity and inferred a causal relationship between sexual reproduction and carotenogenesis. Gametangia and other sexually differentiated structures are normally strongly yellow coloured. Barnett *et al.* (1956) observed that mated cultures of *Choanephora cucurbitarum* produced up to 20-fold more β-carotene per mycelium than single *plus* or *minus* cultures. Some of their results are shown in Table IX. When both mating types were grown on opposite sides of a cellophane membrane, carotenoid synthesis was also stimulated.

TABLE IX. β-Carotene production in mated and single cultures of *Choanephora cucurbitarum* (From Barnett *et al.*, 1956)

Culture	Mycelium dry weight (mg)	β-Carotene (μg)
Plus	25	1 140
Minus	22	1 377
Mated	18	16 560

This observation was extended to other species by several other authors. Sebek and Jäger (1964), Prieto *et al.* (1964) and Sutter and Rafelson (1968) isolated factors from the culture medium of mated *B. trispora* which strongly enhanced carotenoid production in single, especially *minus* cultures of this fungus. These factors were later identified as trisporic acids, as described above. Probably the trisporic acids are the only agents responsible for the stimulation of carotenogenesis observed. Table X gives some data about the effect of natural concentrations of trisporic acids on carotene production in *plus* and *minus* mycelia. They show that it is of the same order of magnitude as is caused by mating. The effect of trisporic acids B and C on individual carotenoids is shown in Table XI. All carotenoid levels go upward, particularly β-carotene.

TABLE X. Influence of trisporic acids on carotenoid production in cultures of *Blakeslea trispora* (from van den Ende, 1968)

Culture	Addition[a]	Carotenoid production[b]
Minus	No	$0\cdot14\pm0\cdot03$
Minus	Yes	$0\cdot73\pm0\cdot22$
Plus	No	$1\cdot05\pm0\cdot07$
Plus	Yes	$1\cdot02\pm0\cdot07$
Mated	No	$0\cdot95\pm0\cdot11$

[a] Trisporic acids were added to a concentration equal to the natural concentration in mated cultures.

[b] Amounts of carotenoids produced are expressed by the optical density at 450 nm of 100 ml of hexane extract per g dry weight. Values are averages of triplicates; maximal deviations from the average are indicated.

Presumably the effect on carotenogenesis is caused by the stimulating influence of trisporic acids on the synthesis of a limiting enzyme involved in the biosynthesis of β-carotene (Thomas and Goodwin, 1967). This is supported by the fact that the action of trisporic acids is inhibited by cycloheximide, an inhibitor of protein synthesis, as is shown in Fig. 13. Cycloheximide must be applied simultaneously with or prior to the application of trisporic acids. If applied several hours later it has no effect on the production rate of β-carotene (Thomas *et al.*, 1967). The enzyme, the synthesis of which is stimulated by trisporic acids, is probably functioning at an early stage in the isoprenoid pathway, since the production of sterols, ubiquinones and prenols is also stimulated by trisporic acids (Bu'Lock and Osagie, 1973).

TABLE XI. Influence of trisporic acids B and C on the production of various carotenoids in *Blakeslea trispora-minus* (from van den Ende, 1968)

Addition	Carotenoids (μg/g dry weight)						
	Phytofluene	β-Carotene	α + β-Zeacarotene	γ-Carotene	Neurosporene	Lycopene
None	—	2	—	1	—	1
Trisporic acid C	2	15	2	13	1	6
Trisporic acid B	9	44	8	20	3	5

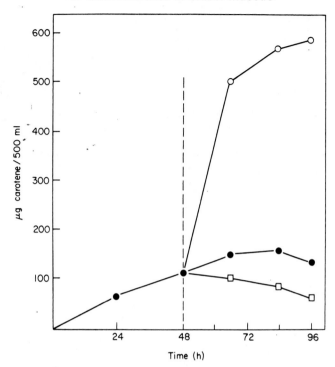

Fig. 13. The effect of cycloheximide on trisporic acid-stimulated carotenogenesis in *Blakeslea trispora minus*. Trisporic acids and cycloheximide (50 μg/ml) were added to 48-h-old cultures which were incubated for a further 48 h. □, Cycloheximide and trisporic acids added; ○, trisporic acids alone added; ●, control (from Thomas *et al.*, 1967).

The question can be asked, what the function is of high carotene levels in sexually active mycelium. At least two possibilities have been envisaged. Gooday and coworkers (1973) have shown that during sexual reproduction β-carotene is oxidatively polymerized to give sporopollenin, a component of the zygospore wall, where is functions to provide protection against both chemical and biological attack. Furthermore, cyclic carotenoids are precursors of the trisporic acids (see next section). The total amount of material derived from the carotene pools for trisporic acid production generally largely surmounts the size of these pools in a mated culture of *Blakeslea trispora*. It was observed, for instance, that 2600 μg/g trisporic acids was produced in 48 h by mated mycelium containing 360 μg/g β- and γ-carotene (averaged over *plus* and *minus* strains; van den Ende *et al.*, 1972, and unpublished work). Thus one can imagine that a functional carotenogenic system is prerequisite for trisporic acid synthesis.

A third action of the trisporic acids concerns some of the enzymes which are involved in its own biosynthesis. This will be discussed in the next section.

BIOSYNTHESIS OF THE TRISPORIC ACIDS

The trisporic acids are derived from cyclic carotenoids. This appears from the following evidence. Austin and coworkers (1969b) demonstrated that radioactive β-carotene, when fed to mated *Blakeslea trispora* cultures, is incorporated into trisporic acids. Inhibition of β-carotene synthesis by diphenylamine also prevents trisporic acid production. Furthermore, caroteneless mutants of *Phycomyces blakesleeanus* do not produce trisporic acids in mated cultures (Bergman *et al.*, 1969; Sutter, 1975). Tracer studies by Bu'Lock and associates (1974a) have provided evidence concerning the sequence of reactions between β-carotene and trisporic acids. In Table XII some compounds are presented which have been shown to contribute label to trisporic acids when added to mated cultures in radioactive form. This suggests that they are precursors of the latter. However, the primary product of the cleavage of β-carotene is unknown.

Any hypothesis which is designed for the pathway of trisporic acid synthesis must take into account the fact that these hormones are only produced by mated cultures. The problem how to explain that full-grown *plus* and *minus* cultures start to produce trisporic acids only after mixing, has provoked several investigations.

The following points are relevant for the solution of this problem. Van den Ende and coworkers (1968, 1970) demonstrated that trisporic acid synthesis also occurs when the two mating types are kept separated by a membrane filter. Even when the barrier consists of two membrane filters separated by a small amount of permanently sterile medium, the synthesis is not prevented; it is only retarded. Another significant result was that trisporic acid synthesis in a mated culture appears to take place in the *plus* as well as in the *minus* part of the culture. When one of the strains is ^{14}C-labelled, the trisporic acids formed are radioactive, irrespective which of either mating type is labelled (van den Ende *et al.*, 1972). Table XIII shows the specific radioactivities of the trisporic acids when the *plus* or the *minus* is radioactive. It appears that the specific activities of the trisporic acids are roughly half of the specific activities of the radioactive mycelia, and, in addition, are more or less constant in time. This indicates

TABLE XII. Incorporation of possible intermediates into trisporic acids by cultures of *Blakeslea trispora* (from Bu'Lock *et al.*, 1974a)

Compound	Spec. radioactivity (d.p.m./mmol)	Incubation time (h)	Spec. radioactivity of trisporic acid C after incubation (d.p.m./mmol)
[10-^{14}C]-retinol	$1 \cdot 3 \times 10^8$	8	$4 \cdot 8 \times 10^6$
[10-^{14}C]-β-C$_{18}$-ketone	$7 \cdot 9 \times 10^7$	16	$7 \cdot 7 \times 10^6$
[10-^{14}C]-4-hydroxy-β-C$_{18}$-ketone	$1 \cdot 8 \times 10^8$	2	$2 \cdot 6 \times 10^7$

TABLE XIII. Specific radioactivities of trisporic acids from combined ^{14}C-labelled *plus* and non-labelled *minus* mycelia of *Blakeslea trispora*, and vice versa (from van den Ende *et al.*, 1972)

Time after combination (h)	Specific radioactivities of trisporic acids (d.p.m./μatC)[a]	
	Plus mycelium labelled (760 d.p.m./μatC)	*Minus* mycelium labelled (1215 d.p.m./μatC)
11·5	368$\pm$33	
22	399$\pm$14	
34	342$\pm$ 3	
46	332$\pm$32	
15·5		524
24		496$\pm$ 0
39		507$\pm$11
50		495$\pm$13

[a] The radioactivities are expressed in d.p.m./μatC ($= \frac{1}{18}$ d.p.m./μmol in the case of trisporic acids) to allow comparison with the radioactivities of the respective mycelia. Values are averages of duplicates; deviations from the average are indicated.

that both mating types contribute about equally to trisporic acid production. Thus in the mated condition both mycelia produce trisporic acids, but when single, neither is active in this respect. The possibility was envisaged that the synthesis of key enzymes, involved in this process, was induced by factors diffusing from one mate to the other. This was supported by the fact that the inhibitors of protein synthesis, 5-fluorouracil and cycloheximide, strongly inhibited trisporic acid synthesis (van den Ende *et al.*, 1970; Bu'Lock and Winstanley, 1971). This concept seemed less tenable, however, when Sutter and coworkers, after careful analysis, discovered that *plus* as well as *minus* mycelia secreted small amounts of neutral metabolites which were converted by the mating partner. Thus trisporic synthesis can take place by way of cross-diffusion of metabolites, necessarily only in mated cultures (Sutter, 1970; Sutter *et al.*, 1973, 1974). Table XIV shows some of Sutter's results.

Some of the metabolites were soon characterized. From extracts of *plus*-culture fluid the 4-hydroxy methyltrisporates, and from *minus* liquid the trisporins were identified as major trisporic acid precursors (Bu'Lock *et al.*, 1974b; Nieuwenhuis and van den Ende, 1975). In addition, the trisporols were found in extracts of *minus* cultures (Austin *et al.*, 1970). These com-

TABLE XIV. Trisporic acid synthesis from metabolites extracted from media of single cultures (from Sutter *et al.*, 1974, modified)

Culture	Extract added	Radioactivity of extract added (c.p.m.)	Radioactivity recovered in trisporic acids (c.p.m.)
Minus	*Plus* extract	$2\cdot35 \times 10^5$	30 160
Plus	*Plus* extract	$2\cdot35 \times 10^5$	830
Minus	*Minus* extract	10^5	320
Plus	*Minus* extract	10^5	18 840

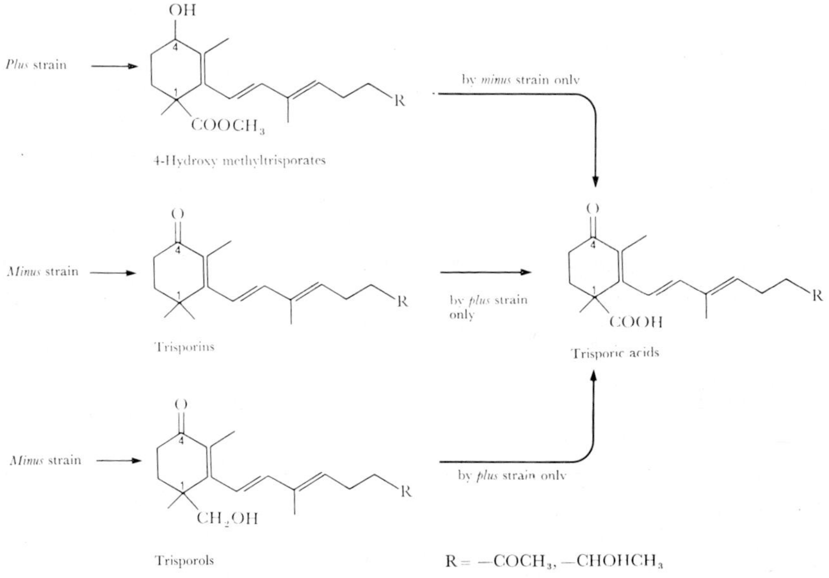

FIG. 14. Sex-specific production and metabolism of precursors of the trisporic acids.

pounds and their conversion reactions are shown in Fig. 14. The 4-hydroxy methyltrisporates are converted to trisporic acids by *minus* strains only. This implies the oxidation of the hydroxy group at C-4 and hydrolysis of the carboxymethyl group at C-1. The trisporins are converted, probably via the trisporols, only in the *plus* strain, by oxidation of one of the *gem*-dimethyl groups at C-1. Apparently *plus* and *minus* strains differ in their ability to carry out the complete oxidation at either C-4 or C-1, respectively, of the trisporic acid skeleton (cf. also Bu'Lock *et al.*, 1973). Trisporic acid synthesis can thus proceed in mated cultures as the result of the combination of two incomplete but complementary synthetic pathways, which at least partly exhibit strong sex specificity. It may be emphasized that these results were obtained in *Mucor mucedo* and *Blakeslea trispora* alike. This confirms that the observed specificity is sex-linked and not based on an accidental difference between non-isogenic strains.

Figure 15 shows the pathway of biosynthesis as it can now be envisaged to occur in mated cultures of *Mucor mucedo* and *Blakeslea trispora*. Some comments must be made about the presented scheme. In the first place, it is uncertain if any mating-type specificity can be assigned to the production of β-C_{18}-ketone and 4-hydroxy-β-C_{18}-ketone (compounds **I** and **II** in Fig. 15), since their conversion into trisporic acids was only demonstrated in mated cultures (Bu'Lock *et al.*, 1974a). Their occurrence *in vivo* was not demonstrated. In the second place it may be remarked that presumably the 4-hydroxy methyltrisporates (**III**) are first oxidized to methyltrisporates (**IV**), which are then hydrolysed to give the trisporic acids as indicated in Fig. 15, since the latter reaction can be easily demonstrated in *minus* mycelia (Bu'Lock *et al.*, 1972). In the third place it may be noted that the production rates of the intermediates described in single cultures is by far insufficient to explain the high rate of trisporic acid synthesis in *B. trispora* (Sutter, 1970; Sutter *et al.*, 1973). For instance, from 15 litres of *plus* culture liquid 1·6 mg 4-hydroxy methyltrisporates was isolated by Bu'Lock *et al.*, whereas their reported yield of trisporic acids amounts to 500 mg trisporic acids from the same volume of mated culture liquid (Bu'Lock *et al.*, 1972, 1974b). Also in *M. mucedo* the observed rate of intermediate synthesis by single cultures is much lower than the rate of trisporic acid formation in corresponding mated cultures. The solution of this problem was offered by the finding that trisporic acids themselves strongly stimulate the production rates of these intermediates, as is shown in Table XV. This stimulation is effectively inhibited by 5-fluorouracil and cycloheximide (Werkman and van den Ende, 1973). Thus trisporic acid synthesis appears to be self-enhancing. Once formed in mated cultures, trisporic acids stimulate their own synthesis, which explains their relatively high production rates.

The exact site in the reaction sequence which is affected by trisporic acids is unknown. The β-carotene levels in single strains, although increased by trisporic acids (see page 62), are presumably non-limiting for precursor synthesis, because also in the presence of glucose (which is

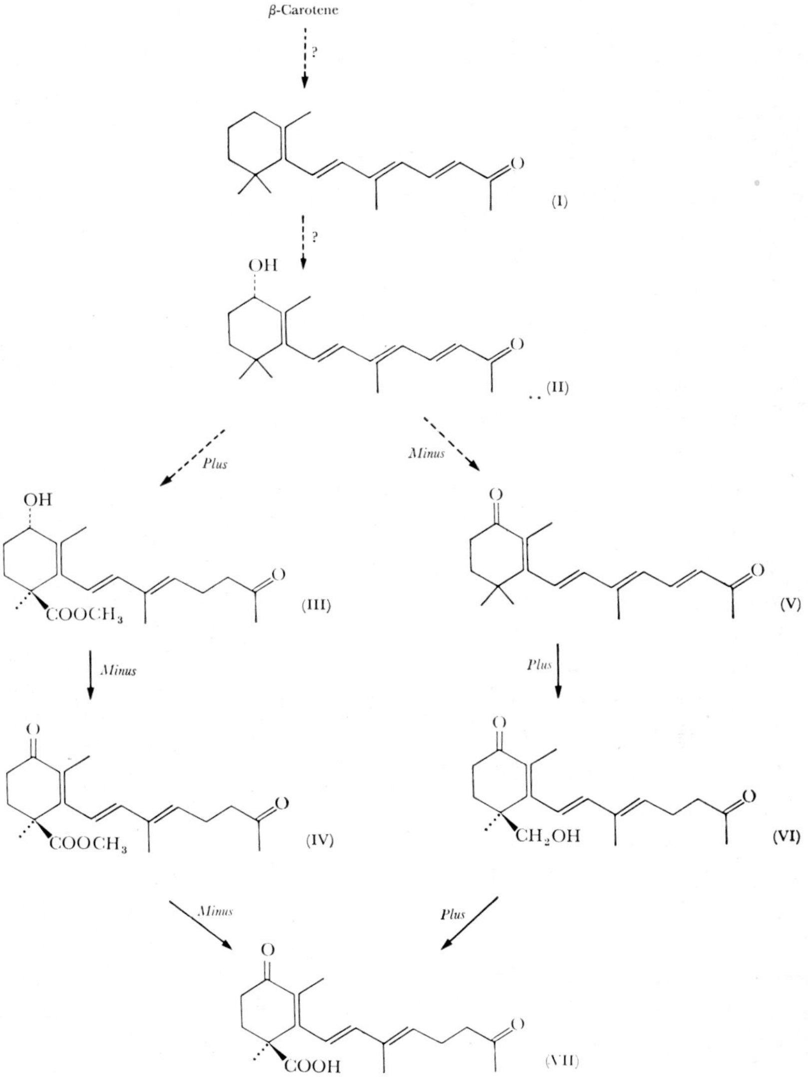

FIG. 15. Proposed biosynthetic pathway of trisporic acid B in mated cultures of *Mucor mucedo* and *Blakeslea trispora*.

TABLE XV. Influence of trisporic acids on neutral precursor production in *plus* and *minus* mycelium of *Mucor mucedo* (from Werkman and van den Ende, 1974)

Mating type	Trisporic acids added[a]	Neutral precursor material produced (units)[b]
Plus	No	69 + 39
Plus	Yes	891 ± 81
Minus	No	207 ± 117
Minus	Yes	1863 ± 1053

[a] The amount of trisporic acids added as 54 μg per Petri dish, containing 20 ml culture.

[b] The amounts of precursor material were estimated by serial dilutions and expressed in units, one unit being defined as the smallest quantity still exhibiting induction of zygophores. Values are averages of duplicates. Deviations from the average are indicated.

reportedly a readily used carbon source for this pathway)* trisporic acids have a stimulating action upon their own synthesis. This invalidates the possibility that the stimulation is by way of enhancing carotenoid synthesis. However, it has recently been observed that the enzyme activities which convert 4-hydroxy methyltrisporates (**III**) to methyl trisporates (**IV**), and the methyl trisporates to trisporic acids (**VII**), are strongly increased by the addition of trisporic acids to the *minus* strain (B. A. Werkman, unpublished work). This is illustrated in Table XVI. If one assumes that the same enzymes are operative at different sites in the reaction sequence, and are specific for a certain region in the substrate molecule, rather than for the molecule as a whole (cf. Bu'lock *et al.*, 1973), one can imagine that their increased activities lead (a) to higher precursor production rates, and (b) to higher conversion rates of these precursors to trisporic acids. Thus the enzyme which is responsible for the hydroxy group at C-4 is active not only in the reaction step between compounds **II** and **V**, but also in that between **III** and **IV** in the *minus* strain. Although the evidence for this double action of an enzyme has at present only been obtained in the *minus* mating type, it is reasonable to assume that the same situation occurs in the *plus* mating type, where one enzyme might be operative

* Note: Presumably, the glucose carbon is channelled into the trisporic acid pathway via the carotenoid pools. This does not contradict the idea that glucose may be a relatively better source of carbon than the carotenoids, because probably only part of these pools are accessible for metabolic processes.

TABLE XVI. Effect of trisporic acids on the enzyme activities converting 4-hydroxy methyltrisporates to trisporic acids in the *minus* strain of *Mucor mucedo* (B. A. Werkman, unpublished work)[a]

Pre-incubation period (h)		Enzyme activity (nmol trisporic acids/h/mg protein)
Cycloheximide	Trisporic acids	
0	0	$11 \cdot 43 \pm 0 \cdot 14$
0	3	$22 \cdot 36 \pm 0 \cdot 17$
0	18	$31 \cdot 23 \pm 0 \cdot 81$
4	0	$3 \cdot 84 \pm 0 \cdot 16$
4	3	$3 \cdot 06 \pm 0 \cdot 42$
19	0	$2 \cdot 48 \pm 0 \cdot 16$
19	18	$2 \cdot 79 \pm 0 \cdot 18$

[a] Five-day-old mycelium was pre-incubated with trisporic acids and/or cycloheximide prior to homogenization of the mycelium. The enzyme activity was determined in the 50 000 *g* supernatant.

not only in the reaction between compounds **II** and **III**, but also between **V** and **VII**, being responsible at both sites for the oxidation of the methyl group at C-1.

The hypothesis elaborated above also explains why frequently a defect in one mating type completely blocks trisporic acid synthesis in the partner at combination. For example, in the combination of a defective *minus* mutant with a wild-type *plus* strain of *Mucor mucedo* the mutation may not only block zygophore development in the mutant but also in its wild-type partner (Wurtz and Jockusch, 1975). The explanation could be that the mutant has a lesion in its trisporin production, i.e. in the oxidation of C-4 of the trisporic acid skeleton. This block would also inhibit the conversion of *plus*-derived 4-hydroxy methyltrisporates. Thus no trisporic acids would be formed at all. A similar explanation can be given for the low trisporic acid production in the combination of a culture which is pretreated with fluorouracil with its untreated mate (Bu'Lock and Winstanley, 1971).

Finally it should be stressed that the physiological and chemical aspects of sexual differentiation are strikingly similar in all but the quantitative aspects in *Blakeslea trispora, Mucor mucedo, Phycomyces blakesleeanus* (Sutter, 1975), and *Absidia glauca* (Uyama, 1972). This explains the ready occurrence of interspecific matings in this group of fungi.

BIOLOGICAL PROPERTIES OF TRISPORIC ACID PRECURSORS

While trisporic acids exhibit no specificity with respect to zygophore induction in *plus* and *minus* strains, the opposite is true for the precursors of these sex factors. The *plus*-derived 4-hydroxy methyltrisporates specifically induce zygophores in the *minus,* and so do the trisporins in the *plus* mating type. It can be argued that these and other precursors are in themselves inactive but exert their apparent function only after conversion to trisporic acids by the appropriate mating partner. However, there is no evidence to support this. There are mutants known which respond to the addition of 4-hydroxy methyltrisporates by producing zygophores, but which are insensitive towards trisporic acids. This suggests a different perception mechanism for trisporic acids and its precursors. Alternatively, these mutants might be blocked in the uptake of trisporic acids, but not in their sensitivity for these compounds, whereas neutral precursors might be able to readily penetrate into the cell and be converted to trisporic acids. The latter supposition has been expressed by Wurtz and Jockusch (1975).

Trisporic acid precursors can induce zygophores via the air. This was unambiguously demonstrated by placing a glass vial with an aqueous solution of these precursors in front of the appropriate mycelium of *Mucor mucedo* (Nieuwenhuis and van den Ende, 1975). This explains the phenomenon reported by Mesland *et al.* (1974) that zygophores are induced when two unlike vegetative mycelia (producing precursors, not trisporic acids) are positioned on either side of a slit in the supporting agar, inhibiting diffusion contact via the substrate (see also page 56). One could thus imagine that the trisporic acid precursors have a predominant function in sexual interaction, more than the trisporic acids, which are the end products, formed endogenously from cross-diffusing intermediates. Clearly, *Blakeslea trispora,* which excretes such large amounts of trisporic acids, is an exception among the numerous species which synthesize trisporic acids on a more modest scale, and largely intracellularly (Gooday, 1968; Sutter *et al.,* 1973).

CHEMOTROPISM

Zygophores of unlike mating type tend to grow towards one another, often curving in order to accomplish contact. It seems clear, therefore, that zygophore growth is somehow directed, and that physical contact between the mating partners is by no means accidental. Burgeff (1924), who reported this phenomenon for the first time, introduced the term "zygotropsim" for this process (cf. page 55).

Banbury (1954) suggested that a stimulus is transmitted by the excretion of volatile substances by the zygophores, causing a positive tropic response in the sexual partner. Plempel and coworkers (Plempel, 1960; Plempel and Dawid, 1961; Plempel, 1962) studied the phenomenon in more detail and were able to confirm Banbury's suggestions. The tropic stimulus is volatile, very sensitive to oxygen and therefore short-lived under natural conditions. Mesland *et al.* (1974) observed that zygophores not only are attracted by zygophores but also by vegetative mycelium and sporangiophores of the other mating type. This suggests that the substances involved are secreted by vegetative mycelium. A neutral extract of culture medium appears to have an orienting effect upon zygophore growth, but the responsible components have not been identified (Mesland *et al.*, 1974).

MONOECIOUS MUCORALES

In these species, sexual differentiation is very similar to the process described earlier in this chapter for dioecious Mucorales. In the relatively well-observed *Zygorhynchus moelleri* a slender side branch originates from a main hypha, usually just below a septum. This side branch generally grows upward with strong curvature towards a higher part of the main branch. After contact the main branch forms a very short lateral extension which bears one of both gametangia (Fig. 16). Invariably, the copulatory side branch is filled with dense cytoplasm and fat globules, contrary to the main branch which contains "normal" cytoplasm. In other monoecious species, sexual phenomena may deviate more or less from this description, but generally it is evident that some part of the mycelium becomes different from another part of the mycelium in sexual respects. This then gives rise to a conjugation process which resembles that in mating dioecious species.

This similarly is accentuated by interspecific conjugations between dioecious and monoecious species. The dioecious partner frequently produces zygophores which grow towards monoecious sexual hyphae, and vice versa. Irregular swellings originate where both mates come into contact and often gametangia are delimited. Fertile zygospores are never formed.

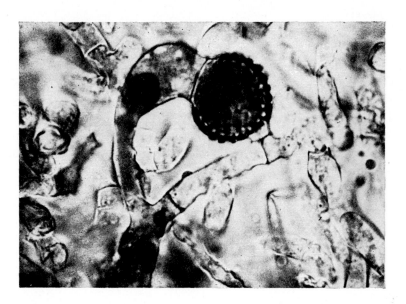

FIG. 16. Conjugation in *Zygorhynchus moelleri* (× 1850).

Satina and Blakeslee (1930) observed that monoecious species do not always react similarly in this type of mating. Some show conjugative behaviour only, or preferably, with *plus* mating types thus exhibiting a *minus* "tendency". An example of this type is *Zygorhynchus moelleri*. Of this species five different isolates were studied, which all had the same preference for dioecious *plus* strains. This suggests that the tendency exhibited is a specific and not a racial property. A *plus* tendency was exhibited by *Z. heterogamus*, whereas *Absidia spinosa* produced sexual structures with *plus* as well as with *minus* strains. Some species, like *Sporodinia grandis* (= *Syzigites megalocarpus;* Zycha *et al.,* 1969) did not show any tendency to conjugate with either mating type. This is more or less reflected by the *plus* or *minus* character of the separate sexual structures. In the case of *Z. moelleri,* only the main hyphae, which are most abundant, copulate with *plus* hyphae in interspecific contrasts, suggesting that they have a *minus* nature. In other species the situation is uncertain since frequently a "sex reversal" seems to take place upon branching (Burgeff, 1924).

The interspecific sexual reactions do suggest that in monoecious species also the hormonal system that has been disclosed in dioecious Mucorales is operative. This is supported by the fact that zygospore formation in

TABLE XVII. Some sexual characteristics of monoecious Mucorales

Species	Character according to Satina and Blakeslee (1930)	Production of known mating factors	Response to trisporic acids	Type of conversion reactions exhibited
Zygorhynchus moelleri	*Minus*	Trisporins, trisporic acids	No	*Minus*
Zygorhynchus heterogamus	*Plus*	Trisporic acids	No	Neutral
Mucor genevensis	Neutral	—	Yes	*Minus*
Syzigites megalocarpus	—	—	Yes	Neutral

seemingly sterile monoecious species can be induced by contact with other monoecious or dioecious species, or by the addition of trisporic acids and its precursors. Schipper (1971) described an agamic strain of *Zygorhynchus moelleri*, which was totally devoid of sexual activity. However, when grown in the presence of the *minus* strain of dioecious *Mucor* species, zygospores were produced near the line of contact of the colonies. Also, when trisporins were applied, zygospores were formed (B. A. Werkman, unpublished work). The same phenomenon was observed in the species *Syzigites megalocarpus.* A strain unable to produce zygospores under standard conditions, did so readily when trisporic acids or its precursors were applied (Werkman and van den Ende, 1974). Also, in *Mucor genevensis* application of these factors had a stimulatory effect on otherwise normal zygospore production. On the basis of these facts it would be easy to assume that trisporic acids and related compounds play a role in sexual reproduction in monoecious species, if only it were possible to demonstrate their presence and biosynthesis. In some respect this was established by Werkman and van den Ende (1974), who demonstrated that some species, listed in Table XVII, are able to convert trisporic acid precursors. Moreover, they isolated biologically active substances from some of these, which had characteristics similar to those formed in dioecious strains. This all leaves little doubt that the hormonal control of sex in monoecious species is closely related to that in dioecious Mucorales.

5. Sexual Reproduction in Yeasts

Yeasts are distinguished from filamentous fungi by a growth pattern
which results in a more or less globose unicellular form. In the majority
of species, propagation of cells takes place by budding; in some yeasts,
like *Schizosaccharomyces,* cells multiply by binary fission. The unicellu-
larity allows the production of synchronous mass cultures. The advantage
of this is that sexual reproduction can be studied in relation to the life cycle
in a population which is much more homogeneous than colonies of
filamentous fungi, in which large differences between various regions can
occur due to age or differentiation (Burnett, 1968).

Recently, much work has been published on sexual differentiation in
yeasts, which has clarified to a great deal the way individual cells of
different sex interact. However, one should be aware of the fact that
yeasts form no taxonomic unity. With respect to their physiology, there is
great variation between the thirty-nine recognized genera. In particular,
the ascosporogenous yeasts and the basidiosporous yeasts differ profoundly
in sexual respect. Therefore, these groups will be described separately.

ASCOSPOROGENOUS YEASTS

In this group of yeasts vegetative diploid cells can, by altering the culture
conditions (e.g. nutrients, temperature, pH) become transformed to asci
containing haploid spores. The maximum number of spores per ascus is
quite variable, but normally four. These spores germinate and give rise
to stable haploid colonies of cells which propagate by budding. Alternatively, these spores may fuse in pairs, forming zygotes in which nuclear

fusion takes place. Conjugation also occurs between haploid cells, and
about this process most information has been gathered. The zygotes resulting from the mating process proliferate mitotically by budding but they
can also, depending on the conditions, undergo meiosis, followed by
ascospore formation (see Fowell, 1969 and Hartwell, 1974, for review
of the yeast cell cycle).

In most species, monoecious as well as dioecious strains have been found.
Mating types of dioecious strains are generally designated a and α, and
are genetically controlled by one locus and two alleles. These alleles can
sometimes mutate from one form to the other (Ahmad, 1965). Mutation
is also the cause of an occasional switch from dioecism to monoecism, or
of complete loss of mating ability.

CONJUGATION IN ASCOSPOROGENOUS YEASTS

The results of Sena *et al.* (1973) may serve to describe some morphological
and physiological aspects of the conjugation process in *Saccharomyces
cerevisiae*. When unbudded haploid cells of exponentially growing cultures

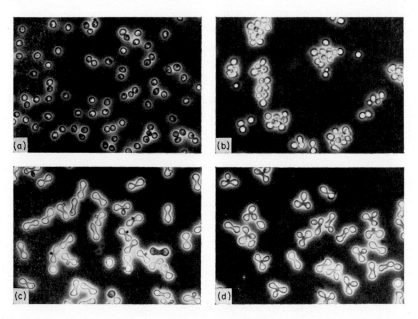

Fig. 17. Sequence of events during synchronous mating in *Saccharomyces cerevisiae*
(a) cells after 20 min; (b) cells after 60 min; (c) purified zygotes, after 120 min;
(d) purified zygotes, after 150 min ($\times 552$) (from Sena *et al.*, 1973).

of opposite mating type are mixed in fresh culture medium, the following events are observed, as is illustrated in Fig. 17. Within 20 min the majority of cells form pairs by cellular adhesion (Fig. 17a). After 60 min marked cell agglutination is observed which is macroscopically visible as a particulate suspension, and microscopically as massive clumping of cells (Fig. 17b). Both cell pairs and agglutinated clumps can be disrupted by sonification. Extensive mating occurs within the agglutinated clumps between 60 and 140 min, resulting in typical dumb-bell shaped zygotes (Fig. 17c). These zygotes appear to form the first bud at the cell junction (Fig. 17d), about 30 min after the onset of cell fusion. The mating is completed within 170 min.

Figure 18 shows the time course of the mating reaction. The most notable feature in this graph is that the cell number per millilitre remains

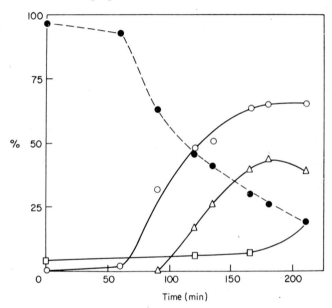

FIG. 18. Time course of the mating reaction in *Saccharomyces cerevisiae*. A mixture of *a* and *α* unbudded cells, isolated from exponentially growing cultures by zonal gradient centrifugation was incubated at 30° C in liquid medium. Samples taken were fixed with formaldehyde, sonified to disrupt agglutinated clumps and counted for budded cells (□–□), unbudded cells (●–●), total zygotes (○–○) and budded zygotes (△–△) by use of a haemocytometer. Cells were scored as zygotes when they appeared fused. The time course of the mating reaction was calculated by defining unmated budded and unbudded single cells as one unit and budded and unbudded zygotes as two units (from Sena *et al.*, 1973).

constant during the sexual process, which means that in this period normal cell division is strongly inhibited. The unmated cells which are left over after completion of conjugation start a round of cell division after 170 min. It is significant that only non-dividing, unbudded cells participate in the conjugation. This was established in the following way. Cultures of single strains were fractionated by sorbitol gradient centrifugation in a zonal rotor so that cells of unequal size were separated to a large extent. Thus fractions enriched in unbudded or budded cells, zygotes etc. were obtained (Halvorson *et al.*, 1971). By testing these separate fractions Sena and coworkers found maximum mating efficiency in samples containing small unbudded cells.

They also found that unfractionated late-log phase or early stationary cultures (in which cell density is high) showed a much reduced zygote production on mixing, compared with younger, mid-log phase cultures (with low cell density). In the former, distinctively malformed, often elongated or pear-shaped cells appeared some time after mixing. They observed the same phenomenon in mid-log phase cultures after increasing the cell density, or after allowing the liquid cultures to stand without shaking.

THE α FACTOR OF *SACCHAROMYCES CEREVISIAE*

The observation that mixing of two opposite mating types in *Saccharomyces* can lead to malformations confirms earlier claims by Levi (1956) and Duntze *et al.* (1970). When, according to these authors, haploid *a* cells are spread on agar near a heavy streak of α cells, the first elongate enormously. After prolonged exposure, the cells attain bizarre shapes that may reach thirty times the size of normal haploid cells. α Cells must be in large excess to evoke this response. This suggests that α cells secrete some factor which affects *a* cells specifically. This was supported by exposing *a* cells to filtrates of α cell cultures. These filtrates appeared not only to cause deformations of cell shape but they also exerted a strong inhibition of cell budding. On this feature further research was concentrated. Duntze and coworkers were able to isolate a factor from culture medium of α cells, acting specifically on *a* cells, causing malformations as well as inhibition of mitosis (Throm and Duntze, 1970; Duntze *et al.*, 1973).

The α factor, as was called the material produced by α cells and acting specifically on *a* cells, was isolated from culture liquid by adsorption on a column of Amberlite CG50 (a weakly acidic ion exchanger), equilibrated with 0·1 N acetic acid. After washing the column with 50% ethanol, the α factor was eluted with 0·01 N HCl in 80% ethanol. The eluate was concentrated and neutralized, and the resulting precipitate removed by centrifugation. The clear supernatant contained the active factor.

The biological activity of the α factor was determined by exposing a cells on agar plates to active fractions and observing the morphological response directly under the microscope after 4–5 h at 30°C (note that the effect on morphology and cell division rate was a priori thought to be caused by the same factor). Solutions to be tested were pipetted into small wells made in the agar. The yeast cells were streaked along the edges

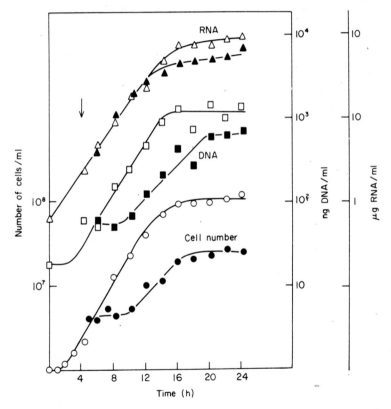

FIG. 19. Inhibition of cell division and DNA synthesis in a cells by the α factor. The factor was added as indicated by the arrow to one portion of the culture at a final concentration of 10 units/ml. Open symbols, control; closed symbols, culture containing α factor; ○, cell number per ml; □, ng/ml DNA; △, μg/ml RNA (from Throm and Duntze, 1970).

of the wells. By testing serial dilutions the concentration was estimated. The lowest concentration causing a detectable response was defined as one unit of activity per millilitre. Total activity was calculated by multiplication of the maximal dilution factor by the total volume of the active fractions.

Subsequent purification by thin-layer chromatography resulted in a biologically active preparation which stained with ninhydrin, indicating the presence of amino acids. Atomic absorption spectrometry showed bands characteristic of copper. The molecular weight was estimated by gel filtration on Sephadex G25 to be between 1000 and 2000.

On the basis of these results the factor is thought to be a peptide, isolated as a copper complex. Whether copper is necessary for the activity of this

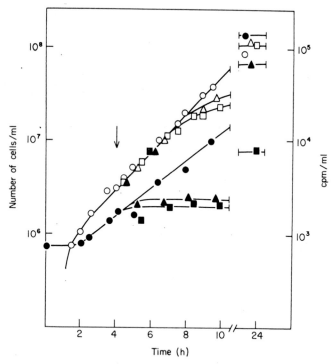

FIG. 20. Incorporation of [^{14}C]leucine in the presence of α factor. Cells were grown in the presence of [^{14}C]leucine; the α factor was added after 4 h (arrow) at concentrations of 10 and 50 units/ml. Cells were precipitated with 5% TCA, incubated at 98° C for 30 min, collected on cellulose nitrate filters, washed and counted. Open symbols, c.p.m. of [^{14}C]leucine incorporated; closed symbols, cell number per ml; ○, control; △, 10 units/ml of α factor; □, 50 units/ml of α factor (from Throm and Duntze, 1970).

factor could not be established, however, since copper is an essential element in every culture medium.

The α factor appeared to have a profound influence on the cell cycle of *a* cells (Throm and Duntze, 1970); Bücking-Throm *et al.,* 1973; Shimoda and Yanagishima, 1973). Figures 19 and 20 show the influence of the α factor on cell division and nucleic acid and protein synthesis in a culture of *a* cells. After addition of the factor, the cells continue to divide at the same rate as the control until a plateau is reached rather abruptly. DNA synthesis is inhibited for several hours, whereas RNA and protein syntheses continue at the same rate as the control culture. The arrest of cell division which is attained after addition of the α factor is released by filtering the cells and transferring them to fresh medium. Approximately 80 min after transfer, a synchronous pulse of cell division is observed, leading to an increase of cell number by a factor 1·7–1·8.

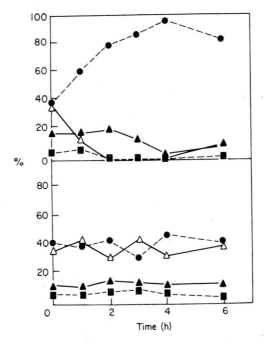

FIG. 21. Cell types as a function of time in a culture of *a* cells exposed to α factor (25 units/ml). One ml samples were removed and fixed with formaldehyde. The fixed samples were stained and scored for: ●, unbudded, mononucleate △, budded mononucleate cells; ■, budded cells with a dividing nucleus; ▲, budded, binucleate cells. Top graph is *plus* α factor and lower graph is control (no α factor added) (from Bücking-Throm *et al.,* 1973).

These results are consistent with the supposition that a cells are inhibited by the factor only during a fraction of the cell cycle, because the large majority of cells accumulate at the same point in this cycle, namely unbudded cells. This is illustrated in Fig. 21 where the percentage of different cell types is examined as a function of time after addition of the α factor. The percentage of budded cells with a single nucleus decreases immediately; the percentage of budded cells with two nuclei remains constant for about 2 h and then begins to decrease; the percentage of unbudded cells increases continuously until it comprises 90% of the population. In an untreated control culture the percentage of each of the four morphological cell types remains constant.

From microscopic analysis it appears that cells with buds at the time

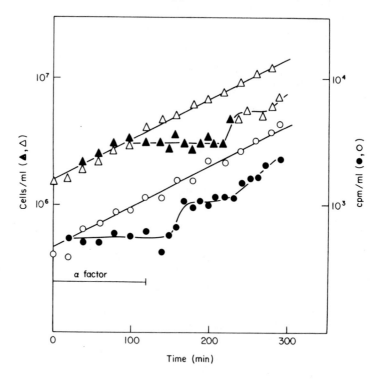

Fig. 22. Synchronization of DNA synthesis and cell division by treatment with α factor (50 units/ml). After 120 min the factor was removed from the treated culture by filtration and the cells were washed and resuspended in the same volume of fresh medium. DNA synthesis was measured by following the incorporation of ^{14}C-adenine into the alkali-resistant, DNase-sensitive material. Open symbols are controls (from Bücking-Throm et al., 1973).

of α factor addition continue to enlarge the bud and form two cells that do not divide further. About one-half of the cells that have no bud at the time of exposure form one in the presence of the α factor and produce two cells, while the other half of the unbudded cells are arrested without forming buds. Thus, there is a period between cell separation and bud emergence during which a cells are responsive to the α factor. Cells that have passed the end of this period will complete their life cycle by forming two cells, while cells that are before this point will remain as unbudded cells, and will accumulate as such.

Since it has been shown that initiation of DNA synthesis occurs just before the emergence of a new bud (Hartwell, 1973), it is reasonable to assume that it is the initiation of DNA synthesis which is inhibited by the α factor. This supposition was supported by the results presented in

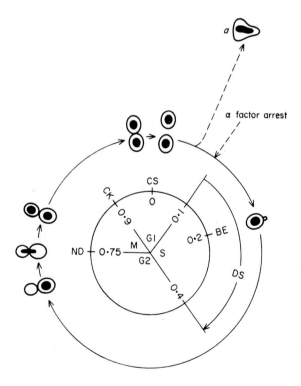

Fig. 23. Summary of the effect of the α factor upon the cell cycle of a cells. Numbers indicate the approximate timing of the following events in the cell cycle: BE, bud emergence; DS, DNA synthesis; ND, nuclear division; CK, cytokinesis; CS, cell separation (from Bücking-Throm *et al.*, 1973).

Fig. 22. This figure shows that DNA replication does not stop abruptly after addition of the α factor. A significant amount of DNA, representing between 20 and 40% of the amount present at the time of α factor addition, is still synthesized. After removal of the factor a synchronous burst of DNA replication takes place which indicates that the initiation of DNA synthesis rather than DNA synthesis itself is affected.

Figure 23 shows the cell cycle of yeasts and the period during which the cells are sensitive to the α factor. This period is limited by the cell separation and the initiation of DNA synthesis. This work has been confirmed by Hereford and Hartwell (1974), who specifically mapped the site of arrest within the sequence of a number of gene functions that lead to the initiation of DNA synthesis.

Although these results show that the α factor interferes with some function required for the initiation of DNA synthesis, it cannot be ruled out that also other functions essential for cell division are affected. In any case, they indicate the existence of a control mechanism aimed at an orderly sequence of events in the sexual interaction between a and α cells. As is evident from the work of Sena and coworkers (1973), mating occurs between unbudded cells only. The secretion of the α factor might contribute to the accumulation of a cell type which can participate in conjugation.

One could predict also that a cells secrete a substance acting specifically upon α cells. Although such a factor has not been isolated, its existence has been suggested by several authors (Bücking-Throm et al., 1973; Bilinski et al., 1973; Hartwell, 1974).

SEXUAL AGGLUTINATION

In the description of the conjugative behaviour of synchronous cultures of Saccharomyces cerevisiae (see above), cell fusion appeared to be preceded by a strong adhesion of cells of opposite mating type, which resulted in pair formation and even massive agglutination of cells which could be observed macroscopically. It is reasonable to suppose that agglutination of cells of different mating type is an important step, increasing the mating frequency by enforcing cell contact. That cell contact could be limiting the mating frequency is clearly demonstrated by the fact that centrifugation of mixed yeast suspensions as a means to promote cell contact between competent haploid cells can result in a 100-fold rise in the yield of zygotes (Jacob, 1962; Bilinski et al., 1973).

The phenomenon of sexual agglutination in yeasts was discovered by Wickerham (1956) after the isolation of a number of haploid cultures of *Hansenula wingei*. He reported: "When these cultures were mixed in pairs, those of the same sex yielded a creamy mass of cells which showed no change in consistency. Pairs of opposite sex agglutinated within a few seconds after the beginning of mixing. The cells set to a solid mass which resisted further mixing and adhered tenaciously to the metal loop by which the cells were stirred. Cells of the two sexes grown separately could be mixed, then rolled between the palms of the hands into a ball of cells so rigid it would stand on a surface without flattening. Agglutinated cells cohered so firmly that they did not disintegrate when placed in a beaker of water." In these cell aggregates conjugating pairs of cells were observed within a short time after mixing, and eventually as much as 80% of the total population had fused. The resulting zygotes gave rise to vegetative diploid cells by budding, accompanied by gradual loss of agglutinability of the cell mass. Diploid cells did not agglutinate.

Some haploid isolates of *Hansenula wingei* and *Saccharomyces cerevisiae* agglutinate directly on mixing, while others take 30 min or longer to start agglutination. The latter strains evidently need some mutual stimulation by the opposite mating type before becoming agglutinative. Also in these strains cell fusion takes place at high frequency only after mass agglutination (Sakai and Yanagishima, 1972).

There are also strains that do not agglutinate at all. Such strains produce less than 2% of zygotes when compared with agglutinative strains (Wickerham, 1956).

Agglutination preceding conjugation has been demonstrated in several other species of the ascosporogenous yeasts, like *Citeromyces matritensis*, *Saccharomyces kluyveri* (Wickerham, 1958), and the fission yeast *Schizosaccharomyces pombe* (Calleja and Johnson, 1971). The latter authors found that only 25–35% of the cell population of *S. pombe* would agglutinate. Therefore they investigated whether conjugation took place exclusively among agglutinated cells. The clumps appeared to be stable to repeated washing in distilled water and to gentle agitation and consequently were readily separable from the free cell population. Neither zygotes nor asci were found in the latter. Thus the conclusion is that in yeasts conjugation is largely dependent upon agglutination. The question can be raised what function is displayed by agglutination: does it merely serve to promote or maintain physical cell contact, or does it also involve specific cell-surface interactions inducing sexual fusion? Two lines of research might contribute to an answer: first, the study of the molecular aspects of sexual agglutination; and second, the investigation of the conjugative behaviour of non-agglutinative strains. Some results will be described in the next two sections.

THE AGGLUTINATION FACTORS OF *HANSENULA WINGEI*

Brock (1959, 1965b) showed that sex-specific and complementary surface components are responsible for agglutination of opposite mating types (strains *5* and *21*) of *Hansenula wingei*. The elements of strain *5* cells were also investigated by Taylor and coworkers (e.g. Taylor and Orton, 1968; reviewed by Crandall and Brock, 1968). From the surface of these cells a factor, which adsorbed to strain *21* cells, was isolated by treatment with snail gut enzyme. It could be eluted again by lowering the pH. Extraction of this so-called *5* factor from strain *5* cells destroyed the agglutinability of these cells. When added to strain *21* cell it caused iso-agglutination, suggesting a multivalency with respect to binding ability at the strain *21* cell surface. This means that each *5*-factor molecule can combine with more than one *21*-type cell which results in the formation of cell clumps. The action of the *5* factor appeared to be not only sex- but also species-specific: cells of agglutinative *Saccharomyces kluyveri,* strains *3* and *26,* were not agglutinated by this factor, suggesting that it could play a role in cell recognition.

The following structural characteristics of the *5* factor were elucidated by Taylor and coworkers: it is a glycoprotein with an approximate molecular weight of 9×10^5. It can be inactivated by disulphide-cleaving agents like 2-mercaptoethanol by which it is dissociated into six or seven inactive components, the smaller of which have a molecular weight of 12 000 ($s^0_{25,w} = 1.7\ S$). The factor can be partly reconstituted after removal of the mercaptoethanal by dialysis. It is estimated that five to six of the small units combine with one large unit of 9×10^5 daltons to form the complete *5* factor. The small component retains a weak specific binding activity to strain *21* cells (Taylor and Orton, 1971). The authors proposed that the *5*-factor molecule consists of a large central core to which small fragments, containing the active sites, are joined by disulphide bridges.

This investigation was continued by Yen and Ballou (1974a and b). They compared the structure of the *5* factor with that of the cell wall matrix, which consists mainly of mannoprotein. Also the factor contains mannose as the main sugar, its total carbohydrate content being approximately 85%. The most striking feature of the protein part in comparison with that of the cell wall is the high percentage of serine. Most of the carbohydrate is linked to serine and threonine through an O-glycosidic linkage. This appears from the fact that incubation with alkali, which destroys this linkage through β-elimination, causes the release of 87% of the total carbohydrate as oligosaccharides with 1–15 sugar units. In cell wall mannan, β-elimination gives quite different results: only 9% of the carbohydrate is released from serine and threonine. Thus, although the agglutination factor and cell wall glycoprotein have similar overall composition, their structures are different. Presumably, as suggested by Yen and Ballou, cell wall mannan exists mainly as long polysaccharide chains, whereas the 5 factor consists of short oligosaccharides linked to a peptide chain via O-glycosidic linkages.

The proteolytic enzyme pronase destroys the active site of the 5 factor, which suggests its proteinaceous nature. But also the hydrolytic action of the enzyme exo-α-mannanase results in a partial and gradual loss of binding activity. The authors suggest that the carbohydrate moiety in the 5-factor functions to stabilize the molecule, and that only after a certain amount of the carbohydrate is destroyed does an effect on biological activity become visible.

Taylor and Orton (1968) have pointed out that there is quite some variability in the total size of the 5 factor, depending on the preparation examined and the method of isolation used. Preparations which are homogeneous with respect to activity are generally quite heterogeneous with respect to size and charge. It has been suggested by Yen and Ballou (1974b) that *in situ* several 5-factor molecules may be linked together to form a larger particle.

Brock and coworkers (Brock, 1965a; Crandall and Brock, 1968) demonstrated that also from the cytoplasm and the external culture medium of strain 5 cells a factor with agglutinative properties similar to the 5 factor could be isolated. The molecular weight of this factor appeared to be much lower than the material isolated from the cell surface of strain 5 cells.

Much less work has been published about the agglutination factor isolated from the cell surface of strain 21 cells by trypsin digestion (Crandall and Brock, 1968). It has an estimated molecular weight of 45 000. It binds specifically to 5-type cells, but does not agglutinate them. Therefore, it is considered to be univalent. It neutralizes the activity of the 5 factor. This has led to the prediction that 5 and 21 factors form a complex,

D

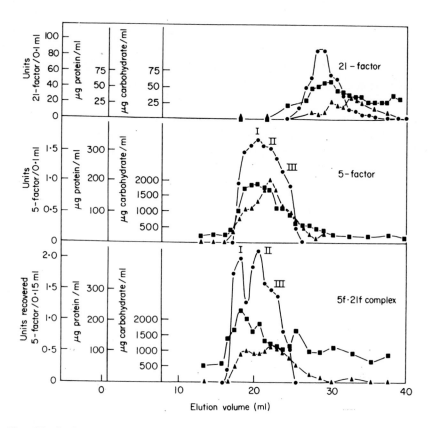

Fig. 24. Sepharose 6B gel filtration of *5* factor, *21* factor and the *5–21* complex, isolated from *Hansenula wingei*. Protein (■), carbohydrate (▲), and biological activity (●) were measured in each fraction. For the *5–21* complex, aliquots of each fraction were first assayed for *5*-factor activity; none was found. Then duplicate aliquots were adjusted to pH 11·5, incubated at 30° C for 30 min, neutralized, and again assayed for *5*-factor activity. Biological activity was assayed by turbidometrically measuring the degree of agglutination in a *21*-type cell suspension after addition of the test solution (from Crandall *et al.*, 1974).

which in fact was proved by Crandall and coworkers (1974). To identify the complex in column fractions, they used a biological assay that is based on the fact that the *21* factor is rapidly inactivated by alkali, whereas the *5* factor is not affected. By treating the inactive complex with alkali, biological activity of the *5* factor was restored. Figure 24 shows the elution profiles on Sepharose of the *5* factor, the *21* factor and the complex, formed by titrating a *5* factor preparation with the *21* factor until agglutination

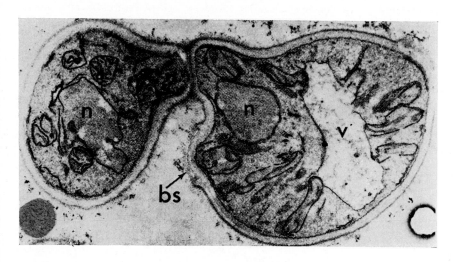

FIG. 25. Section of two cells in an early stage of conjugation. BS, bud scar; N nucleus; V, vacuole (from Conti and Brock, 1965).

activity was just neutralized. The complex was found to be eluted at the same rate at the 5 factor, showing a similar molecular weight heterogeneity.

From diploid cells neither factor has been isolated. Crandall and Brock (1968) have proposed a model of mutual repression in which each haploid carries a structural gene for its agglutination factor and a repressor gene for the complementary factor. Thus, in diploids both factors would be repressed. Interestingly, Crandall discovered that in stationary cultures diploids can become agglutinative with 21-type cells. For this transition from neutral to 5 character (reflecting the synthesis of 5 factor) vanadium salts are thought to be responsible. On the other hand, synthesis of 21 factor by diploids is induced by adding a chelating agent, suggesting that the latter is inhibited by metal ions (Crandall and Caulton, 1973).

The first consequence of sexual cell contact is the formation of a conjugation tube, an extension of the cell strictly in the region where the two cells touch (Fig. 25). This suggests that functional elements, localized at the cell surface, and activated only by sexual contact, are responsible for this phenomenon. Further evidence for this supposition is acquired by preparing agar blocks of conjugation medium and mixed cells at such densities that many of them are in very close proximity but generally do not touch. Only when two cells are actually touching does conjugation occur. If the cells of one mating type are inactivated by ultraviolet light before mixing, no cell fusion occurs. However, in the region of cell contact the non-irradiated cells form protuberances, indicating that even the surface of inactivated cells can evoke a response (Brock, 1965a).

SEXUAL BEHAVIOUR OF NON-AGGLUTINATIVE STRAINS

The question can be asked whether the agglutination factors described above play a role in this surface activation process. This does not seem to be very probable, however, if one considers the results obtained with

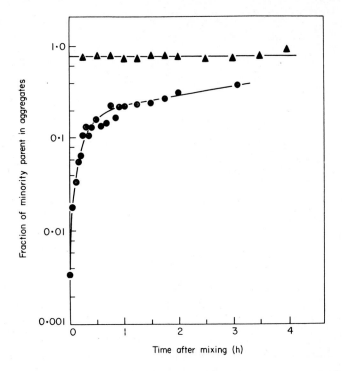

FIG. 26. Kinetics of appearance of diploid colonies under two experimental conditions in non-agglutinative yeasts. The mating efficiency (production of diploids) is plotted as a function of time after mixing in 0·15 M NaCl. The upper curve (▲) shows results obtained from sedimentation of the mating mixture by centrifugation· The lower curve (●) shows results obtained from gently shaken liquid suspension. Samples were taken at various times and plated on nutrient agar allowing zygote formation. Centrifuged mixtures were gently resuspended just before sampling. The mating efficiency was determined by using complementary nutritional mutants which grow poorly as haploids but abundantly as diploids on the used substrate (from Campbell, 1973).

non-agglutinative strains. When such strains are mixed the mating frequency is very low, as stated above. However, this can be much improved if the mixture is centrifuged, which increases the number of cell contacts. Campbell (1973) demonstrated that in non-agglutinative strains cell fusion is preceded by pair formation. Centrifugation of a cell mixture in a medium preventing zygote formation (0·15 M NaCl), followed by resuspension and plating on complete nutrient agar enhances zygote yield dramatically, as is shown in Fig. 26. In centrifuged samples the mating efficiency is about 80% and is independent of the time of prior incubation in the packed (i.e. centrifuged) state. However, in agitated, non-packed mixtures in 0·15 M NaCl, the mating efficiency of samples after plating on nutrient agar increases with time. There is an initial rapid increase in the efficiency of mating, followed by a slower increase. The shape of the curve can be explained by assuming that random collisions between cells of opposite sex lead to the formation of pairs or aggregates, which are not disintegrated on gentle agitation. Pair formation is highly stimulated by centrifugation; on the other hand, vigorous agitation of the mixture reduces the number of zygotes by 90%, indicating that cell adhesion is weak, in contrast to those strains that undergo mass agglutination as was described earlier. Evidently, also in non-agglutinative strains sexual cell contact plays a major role in effecting cell fusion. On the other hand, Campbell's results also suggest that the agglutination factors, described above, might not be essential components of the mating reaction.

CHEMOTROPISM

Finally a peculiar phenomenon must be mentioned which was observed by Herman (1971) in several species of ascosporogenous yeasts. When two cells of opposite mating type are positioned on a solid medium some 15 μm apart, a typical growth pattern can be observed which strongly suggests a positive chemotropism towards the mating partner. Cells of *Hansenula wingei,* strains *5* and *21,* for instance, both form buds which are preferentially aligned in the direction towards the cell or clone of the opposite mating type. The growth pattern becomes evident after 2–3 h. Cells of identical mating type display a negative reaction; new buds are directed away from cells of the neighbouring colony. The response of two different cells to the presence of one another is confined to an orientation of bud initiation. No marked differences in cell size or rate of bud formation between both mating types have been recorded for this species, although *Saccharomyces cerevisiae* reacts slightly differently. These observations suggest that yeast cells produce diffusible substances which sex-specifically orient bud formation.

SUMMARY

The reader may be aware that this chapter is based upon sometimes unrelated reports produced by different workers. It is dangerous, therefore, to present a seemingly coherent picture which pretends to reflect the reality. Nevertheless, a few tentative conclusions can be drawn about sexual reproduction in ascosporogenous yeasts. (1) Any cell which is in a sexually competent, i.e. non-budding, stage is ready to mate. No specialized fusion cells differentiate prior to cell fusion. (2) In liquid cultures the frequency of cell contacts between compatible strains after mixing determines the mating efficiency. Conjugation tube formation followed by cell fusion is triggered by cell contact only. (3) The frequency of effective cell contacts is enhanced by several types of cell interactions in which diffusible and immobile sex-specific compounds play a role. In the first place, cell growth is arrested at that phase of the life cycle where mating is possible. At least in one sexual strain of *Saccharomyces cerevisiae* a diffusible factor which is responsible for this arrest has been isolated. Secondly, cell contact in liquid cultures is enhanced and prolonged by mutual adhesion effected by factors localized at the cell surface. In *Hansenula wingei* these factors have been well characterized. In some strains the production of these agglutination factors is inducible. Soluble factors, involved in the induction of agglutinability have not yet been isolated. Finally, on solid substrates chemotropism seems to be involved in the promotion of cell contact. Nothing is known about the background of this phenomenon of oriented budding.

The events which follow cell contact are unknown. Presumably the cell surface contains elements which play a role in cell recognition and which direct cytoplasmic activities towards the site of contact, resulting in localized cell wall synthesis and cell wall dissolution.

Apart from the opportunities for further study of sexual interactions which the yeasts can offer, they display some notable features in their sexual physiology which make their study particularly relevant; the interaction between complementary cells, by means of diffusible substances through the incubation liquid or by means of cell surface components, interferes with a vegetative cell cycle which can be better synchronized and defined than in filamentous fungi. Research into the control of the cell cycle might also benefit from the natural means described above.

BASIDIOSPOROUS YEASTS

Of this group of fungi, the sexual physiology of the genus *Tremella* has been most studied. The species of this genus produces basidiospores on dikaryotic hyphae. In contrast, the homokaryotic phase that originates from the basidiospores, has a yeast-like form, and propagates by budding. Fusion of two compatible cells leads again to dikaryotic mycelium. Bandoni (1963) established that the mating system of *Tremella mesenterica* is of a tetrapolar type, with two alleles (*A* and *a*) at one locus and a undetermined number of alleles at a second locus (B). Production of dikaryotic mycelium requires that both sets of factors are unlike.

When two compatible strains of *Tremella mesenterica* are streaked in close proximity on a solid medium, many cells are seen to form long hyphal-like tubes. These can be considered as conjugating tubes, since cell fusion occurs at the apices where two tubes of compatible cells meet. Tube formation evidently is caused by diffusible substances because the phenomenon is also observed when two compatible strains are separated by a dialysis membrane (Bandoni, 1965). The active material is found not only to be dialysable, but also resistant to autoclaving. It is produced by each mating type in the absence of the other.

When two compatible isolates are present on one plate, the conjugation tubes appear to grow in a random fashion. Nevertheless, Bandoni (1965) considers it probable that a chemotropic response is involved, since the tubes regularly meet tip to tip, and often seem to have curved near the point of contact.

Conjugation tubes do not grow on forever. When contact with another tube does not occur, hyphal growth is arrested, and buds or conidia are formed near the tips. The final length of the tubes seems to correspond to the amount of inducing substance that is present.

Reid (1974) devised a bioassay for the conjugation tube inducing factor produced by *A* cells. Serial dilutions of the test solutions and the standard solution with culture medium were autoclaved in test tubes and after cooling inoculated with *a* cells. The tubes were incubated overnight and the fraction of the cells with conjugation tubes determined. He found that the *A* factor was most efficiently extracted by butanol. By silica gel chromatography the factor could be separated into three active fractions. Their constitution was not established.

There are parallels with regard to sexual interaction between *Tremella* and some smut fungi, like *Rhodosporidium* and *Ustilago* (Bauch, 1925). For example, many Ustilaginales display dimorphism as described for *Tremella*. Haploid yeast-like cells (called spores or sporidia) proliferate by budding, and when two mating types are together in one culture, conjugation tubes are formed. Tubes reach each other with remarkable precision over distances which may be as great as 50 μm (Bauch, 1925). Although one might expect that in these organsims diffusible substances would also play a role, initiating and orienting hyphal growth, there is no evidence for their existence. On the contrary, Poon *et al.* (1974) have taken time-lapse microphotographs of conjugating sporidia of *Ustilago violacea* and observed that the cells first form pairs, in a manner reminiscent of *Saccharomyces* and only then form conjugation tubes which elongate considerably and push the cells apart.

Inhibitor studies show that the formation of conjugation tubes is dependent on prior RNA and protein synthesis during 3–4 h after mixing of the partners (Cummins and Day, 1974). Conjugation is also prevented when either partner is inactivated by u.v. light during the first 3 h (Day and Cummins, 1974). These experiments indicate that there is a mutual exchange of information between the mating types prior to the morphological changes, which acts at the level of the genome. The nature of the information exchange is unknown which, as in ascomycetous yeasts, seems to depend on cell-to-cell contact. The discovery of hairs at the cell surface ("fimbriae") which are in contact with the plasma membrane through the cell wall is suggested to provide a means of intercellular communication (Poon and Day, 1974).

As in *Saccharomyces,* the mating frequency in *Ustilago violacea* is dependent upon the life cycle since cells of one mating type (with an *a,* allele) are active only in the non-budding stage. In cells of the opposite mating type of this bipolar fungus (with an a_2 allele) the conjugation rate is constant during the whole life cycle (Cummins and Day, 1973).

6. Sexual Reproduction in the Green Alga *Chlamydomonas*

Chlamydomonas is a biflagellate and unicellular green alga. In the genus many types of sexual reproduction are found, ranging from isogamy to oogamy (cf. Wiese, 1969). This chapter will be primarily concerned with the following, best studied, species: *Chlamydomonas moewusii, C. eugametos* and *C. reinhardi.* These species are dioecious and isogamous. Sexual reproduction occurs between motile *plus* and *minus* gametes (sometimes designated as male and female, respectively), which are morphologically identical. The three mentioned species are also hologamous. This means that prior to sexual reproduction whole cells turn into gametes which in morphological respects cannot be distinguished from vegetative individuals.

Gametogenesis is accomplished in vegetative cultures by depletion of the nitrogen source in the culture medium. Readdition of nitrogen at any time blocks the transformation; gametes may even dedifferentiate to vegetative cells. Schmeisser and coworkers (1973) have shown that the time necessary to attain mating competency after nitrogen deprival is dependent upon certain growth parameters of the vegetative cells. For instance, gametic induction by flooding aged cultures of *Chlamydomas reinhardi* on agar plates is established within 2 h, as is shown in Fig. 27, while synchronously grown cells in liquid cultures require as much as 19 h to produce gametes. The reason for the dependence of gametogenesis upon the previous growth history of the cells is not clear.

COURSE OF THE COPULATION PROCESS

The copulation process between gametes of different mating type proceeds according to a two-step pattern which is also found in several other

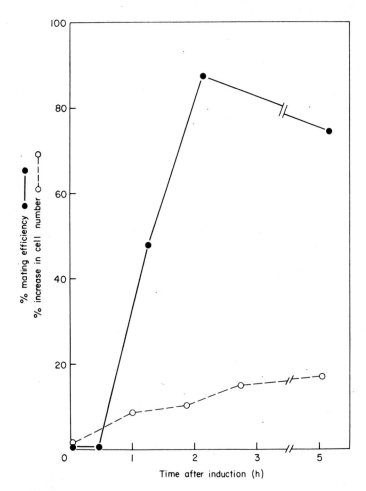

F IG. 27. Mating efficiency and increase in cell number in gametic cultures derived from vegetative cultures of *Chlamydomonas reinhardi* on agar plates. Cells were scraped from an agar surface and suspended in nitrogen-free minimal medium. Equal numbers of cells from *plus* and *minus* cultures were mixed and tested for mating efficiency, 15 min after mixing (from Schmeisser *et al.*, 1973).

isogamous green algae. The first visible stage after mixing of two compatible strains is the adhesion of the mating partners by means of their flagella, in which initially mainly the flagellar tips are involved. In thick suspensions this can lead to the formation of large clumps in which numerous cells of opposite mating type are agglutinated (Fig. 28). Vegetative cells do not take part in this adhesion reaction, indicating that the differ-

FIG. 28. Scanning electron micrographs of *Chlamydomonas eugametos*. (a) mixed *plus* and *minus* gametes involved in sexual agglutination; ×1440; (b) detail showing a *vis-à-vis* pair (arrow) as is evident from the fact that the flagella have become detached from each other; ×7200. The photographs were taken by D. Mesland (unpublished).

entiation of vegetative cells to gametes implies a change at the flagellar surface which results in a highly specific affinity for the flagella of the mating partner.

This sexual adhesion is presumably of a reversible nature. This can be concluded from the observation that flagellar tips frequently change position and that pairs of gametes can be temporarily connected with other pairs (Brown et al., 1968).

No ultrastructural difference between flagella of vegetative cells, gametes or zygotes has been reported. As can be seen in thin sections, they are wrapped in a membrane which is continuous with the plasma membrane. Exterior to the membrane a sheath is present which is visible as an opaque line in electron micrographs, spaced about 20 nm from the flagellar membrane. Its surface shows no defined ultrastructure. Shadowed preparations of whole flagella show fine hair-like projections extending from the flagellar surface. They are called *mastigonemes* (Ringo, 1967; Brown et al., 1968). In *Chlamydomonas moewusii* the flagellar adhesion involves the progressive asociation (or fusion) of the individual flagellar sheaths, beginning at the flagellar tips. Some electron micrographs of thin sections, shown by Brown and coworkers (1968), suggest that this association might even involve the fusion of the flagellar membrane, but this point has not been thoroughly investigated. In any case, this association is very temporary, because, as described below, the flagella separate again when two partner cells have fused.

In *Chlamydomonas eugametos* and *C. moewusii*, which both carry a cell wall during the gamete stage, the following events have been observed after flagellar adhesion. In both partner cells the protoplast forms a short protrusion, also called the plasma papilla, penetrating the cell wall between the points of insertion of the flagella. At the same time, as a result of the pairing of the rapidly moving flagella, both cells come into the correct juxtaposition to establish contact between both plasma papillae. Consecutive fusion of these papillae leads to a protoplasmic bridge between the two partner cells (Brown et al., 1968). The resulting *"vis-à-vis* pair" behaves as one physiological unit. The flagella lose their adhesive properties and while one pair of them resumes its swimming beat, the *vis-à-vis* pair leaves the cluster of agglutinated gametes and moves about for a considerable period of time. Finally, after 18–36 h, the gametes fuse completely and form a diploid zygote. This process lasts only some minutes and involves the broadening of the site of gamete union at the region of the connecting strand. It is followed by the formation of a thick zygotic wall which replaces the remnants of the gametic envelope and by fusion of the nuclei and plastids.

In *Chlamydomonas reinhardi,* which produces naked gametes, the copulation process proceeds slightly different. Also in this species, the cells become united by means of papillae during flagellar adhesion, but the *vis-à-vis* pair stage is very short-lived. The protoplasmic connection widens quickly while the cells allign laterally, until both protoplasts have completely fused to a more or less spherical unit. Only then the four flagella become coordinated in their movement and the zygote starts to swim (Friedman *et al.,* 1968).

THE MECHANISM OF FLAGELLAR ADHESIONS

Since most of the work on sexual reproduction in Chlamydomonads have been concentrated on flagellar adhesion, the attention will be primarily focused on this process.

The preparation of a pure gamete suspension needed for the study of flagellar adhesion is quite straightforward. *Chlamydomonas* species are cultivated on a solid medium, containing salts, in a light regime of 16 h light vs. 8 h dark. After 3–4 weeks the cultures are at a suitable stage to be converted to gametes. This is accomplished quantitatively by flooding the slants with distilled water (cf. Fig. 27). The resulting gamete suspension is washed by centrifugation at 180 g, after which the pellet is taken up in an appropriate buffer (Wiese, 1965).

Förster and Wiese (1954, 1955) observed that the supernatant of a gamete suspension of one mating type after 36–48 h contained material which caused agglutination when added to a suspension of the opposite mating type. The phenomenon appeared to be reciprocal with respect to both mating types of *Chlamydomonas eugametos, C. moewusii, C. reinhardi,* and another not-identified species. The material was sedimentable at 25 000 g. Its susceptibility for proteolytic enzymes and its sugar content led to the suggestion that a glycoprotein was involved (Förster *et al.,* 1956).

The agglutinative action of this material was explained by Förster and coworkers by assuming that it adsorbed to the flagellar surface in a sex-specific way and, owing to its multivalency, would cause isoagglutination of gametes of the same mating type. The authors suggested that this isoagglutinating material might be actively involved in normal sexual adhesion of the flagella of gametes, since it displayed exactly the same characteristics. The following results support this suggestion. (1) Iso-

TABLE XVIII. Sexual compatibility and specificity of isoagglutination in four species of *Chlamydomonas*.[a] Extracts of species and mating types on the left column were tested against the mating types and species of the head column (from Wiese, 1965, modified)

		C. eugametos		C. moewusii		C. species		C. reinhardi	
		plus	minus	plus	minus	plus	minus	plus	minus
C. eugametos	plus	—	+	—	+	—	—	—	—
	minus	+	—	+	—	—	—	—	—
C. moewusii	plus	—	+	—	+	—	—	—	—
	minus	+	—	+	—	—	—	—	—
C. species	plus	—	—	—	—	—	+	—	—
	minus	—	—	—	—	+	—	—	—
C. reinhardi	plus	—	—	—	—	—	—	—	+
	minus	—	—	—	—	—	—	+	—

[a] +, Isoagglutination and sexual compatibility between gametes; −, no iso-agglutination, no sexual reaction.

agglutination was only observed with material and gametes derived from sexually compatible mating types of one species. Thus species specificity and sexual compatibility were fully reflected in the isoagglutination reaction as is shown in Table XVIII. (2) Only the flagella were involved in isoagglutination, so it is reasonable to assume that the material which caused isoagglutination adsorbed only at the flagellar surface. (3) Only gametes took part in isoagglutination, vegetative cells not being affected by material isolated from gamete suspension of the opposite mating type. Evidently, no aspecific adsorption occurred.

The phenomenon of isoagglutination, caused by multivalent glyco-protein material released by one mating type and acting upon the other, is reminiscent of the agglutination process in ascosporogenous yeasts. (Chapter 5). Thus one could speculate that as in yeasts agglutination factors, located at the surface of the flagella, would be responsible for the adhesion by forming intercellular complexes. However, the formation of a complex between *plus* and *minus* agglutinating material *in vitro*, similar to that of the agglutination factors of *Hansenula wingei* by Cran-dall *et al.* (1974), was not reported.

The release of active agglutinating material from gametes into the surrounding medium is somewhat puzzling. One could imagine that a significant amount of it in solution would cause isoagglutination when mixing two gamete suspensions of opposite mating types, which might have an adverse effect on the frequency of zygote formation. It could indicate an active secretion by the cell, reflecting a considerable turnover of the cell surface. This is in accordance with the finding by McLean and Brown (1974) that the mating ability of gametes, being destroyed by trypsin treatment, is fully restored after 45 min.

Another piece of evidence, suggesting that the isoagglutinating material which is washed off from gametes is involved in sexual flagellar adhesion, was provided by the action of α-mannosidase on gametes of *Chlamydomonas moewusii* and *C. eugametos*. Only *plus* gametes proved to be susceptible to this enzyme, their ability to enter a sexual reaction with *minus* gametes being rapidly destroyed. *Minus* gametes, on the other hand, were completely resistant. This was corroborated by the action of α-mannosidase on the isoagglutinating material isolated from these gametes. The material derived from *plus* gametes was rapidly inactivated by this enzyme, whereas the *minus* material appeared to be active, even after prolonged exposures (Wiese and Hayward, 1972). It appears then that in *plus* gametes α-mannosidase-sensitive residues play an essential role in flagellar adhesion. It must be noted that *C. reinhardi* is an exception. Neither of its gametes was sensitive towards α-mannosidase.

The presence at the flagellar surface of proteinaceous material involved in sexual adhesion led Wiese and Metz (1969) to investigate the action of the proteolytic enzyme trypsin. This enzyme was chosen because its pH optimum coincides with that of sexual agglutinability. It appeared that a 0·1% trypsin concentration, which did not interfere with the locomotion of the gametes, seriously affected sexual adhesion. This action was counteracted by trypsin inhibitor. In the mixture of gametes of opposite mating type, incubated with 0·1% trypsin at 26°C, the initial intensive clumping gradually decreased and was finally lost entirely after 45–60 min. No pairs were formed; pair formation also appeared to be very sensitive to the action of trypsin. Separate pre-incubation of the two gamete types and mutual checking with untreated test gametes after addition of trypsin inhibitor demonstrated that in *Chlamydomonas moewusii* and *C. eugametos* only the *minus* mating type was sensitive. In *C. reinhardi*, in contrast, both mating types were sensitive (Wiese and Hayward, 1972). Figure 29 shows the effect of preincubation with trypsin on gametes of *C. moewusii*. It appeared that a minimum time of approximately 25 min was necessary to inactivate *minus* gametes with trypsin, provided that the action of trypsin on pairing was completely neutralized by trypsin inhibitor (which was established in a control experiment). The result can

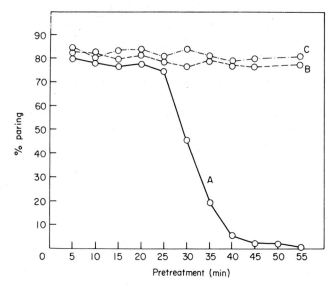

Fig. 29. Trypsin activity of flagellar adhesion in *Chlamydomonas moewusii*. A, A suspension of *minus* gametes was incubated with 0·1% trypsin. At 5 min intervals, 5 ml samples were taken and 5 ml 0·1% trypsin inhibitor added. After addition of 10 ml untreated *plus* gametes the number of *vis-à-vis* pairs was determined, fixing each sample after 30 min. B, Trypsin effect on *plus* gametes checked correspondingly with untreated *minus* gametes. C, Control (from Wiese and Metz, 1969).

be explained by assuming that a substance at the surface of *minus* flagella, but not of *plus* flagella, was removed or inactivated by trypsin (cf. McLean and Brown, 1974).

It should be noted that *in vitro* both isoagglutinating factors were sensitive to trypsin, which the authors explained by assuming that the components *in situ* might be less susceptible to tryptic attack. Moreover, they argued, the capacity to effect isoagglutination requires a functional bi- or multivalent structure, whereas flagellar adhesion presumably could result from interaction of univalent complementary cell surface substances. Accordingly, loss of the multivalent structure would destroy the agglutinating action of the substances in solution, but not necessarily the adhesive material at the cell surface.

In *Chlamydomonas moewusii* and *C. eugametos* isoagglutination of gametes of the same mating type could also be obtained by adding 0·01% concanavalin A (Con A), a haemagglutinin of the jack bean which binds specifically to α-D-glucopyranosyl and α-D-mannopyranosyl residues (e.g. Hassing *et al.*, 1971). Also in this type of agglutination the flagellar tips were involved. However, the action of this substance was not specific with respect to species or mating type (Wiese and Shoemaker, 1970). Vegetative cells were not agglutinated by Con A (McLean and Brown, 1974).

When gametes of *Chlamydomonas eugametos* and *C. moewusii* were incubated with lower concentrations of Con A (0·001%), hardly any isoagglutination occurred. When these incubated gametes after being washed were combined with untreated gametes of the opposite mating type, a clear distinction between *plus* and *minus* gametes became visible. Treated *plus* gametes were no longer able to agglutinate with untreated *minus* gametes, contrary to treated *minus* gametes. These facts suggest that Con A adsorbs to the flagellar surface, inhibiting somehow sexual adhesion of the flagella. *Plus* gametes seemed to be more susceptible than *minus* gametes.

Interestingly, the isoagglutinating material washed off from gametes was also affected by Con A, in the same fashion as gametes were influenced. This material was precipitated from solution by 0·1% Con A. The resulting complexes acted differently: the complex between Con A and *plus* material had no isoagglutinative properties, whereas the complex with *minus* material was still able to agglutinate *plus* gametes. Thus the effect of Con A on this isolated material was paralleled by its action on sexual agglutination *in situ*. This is extra evidence in favour of the hypothesis of Förster and coworkers that the isoagglutinating material which is released by gametes is involved in sexual flagellar adhesion. It is not clear, however, why also *minus* cells were affected by high concentrations of Con A.

Although Con A-mediated agglutination paralleled in some degree sexual agglutination, it is improbable that the same adhesion sites were involved in these phenomena. McLean and Brown (1974) showed that trypsin treatment blocks mating ability but not Con A-mediated agglutinability. Moreover, sucrose and mannose inhibit the Con A agglutination reaction but have no effect on flagellar adhesion. Finally, monovalent Con A, obtained by incubation with trypsin, selectively binds and covers the Con A sites at the flagellar surface without affecting sexual agglutination.

From these discussions it appears that the surface components of the flagellum, responsible for flagellar adhesion, are extremely specific in their action. This is partly reflected by the susceptibilities towards certain enzymes which differ characteristically between *plus* and *minus* gametes.

From these susceptibilities it can furthermore be inferred that in *Chlamydomonas eugametos* and *C. moewusii* important elements of the flagellar surface of *plus* gametes are of a carbohydrate nature. In *minus* gametes of these species, and in gametes of both mating types of *C. reinhardi* protein moieties are also important. In addition, the results indicate that the isoagglutinating substances which are continuously released by *plus* and *minus* gametes are part of the sexual adhesion system.

NATURE OF THE AGGLUTINATION FACTORS OF *CHLAMYDOMONAS*

McLean and coworkers (1974) have tried to elucidate the nature of this agglutinating material by reinvestigating the 25 000 *g* fraction of the culture medium of *Chlamydomonas moewusii* which, according to Förster *et al.* (1956), contains the isoagglutinating activity. After negative staining with phosphotungstic acid in 0·4% sucrose they observed by electron microscopy that this fraction contained membrane vesicles with a fussy coat and a few mastigonemes (cf. page 100) attached. The mastigonemes were 8–10 nm in diameter and 0·4 μm long. The vesicles varied in size and shape and could be as large as 0·5 μm. The fussy coat on the vesicles was 17–34 nm thick. There were no morphological differences between either the mastigonemes or vesicles of the *plus* and *minus* mating types. This fraction was purified by ultracentrifugation in 2·8 M CsCl, which resulted in a fraction containing only vesicles, with strong isoagglutinating activity. Isolated mastigonemes showed no activity. Observation of sections of the biologically active fraction confirmed the membrane nature of the vesicles, the existence of the fussy coat, and the purity of the fraction.

These results suggest that the moieties responsible for agglutination are located at the outer surface of membrane vesicles that bud off the flagellar membrane into the surrounding medium. Their nature remains obscure, however. It is worth noting that vesicle production by gametes is not a necessary element of the mating reaction, since this phenomenon is also observed in vegetative cells of *C. reinhardi* (McLean *et al.*, 1974).

What is the nature of the interaction between *plus* and *minus* flagella during sexual agglutination Any hypothesis should take into account the high specificity and also the transitory character of the interaction. In Chapter 1 a proposal by Roseman (1970) was described, stating that glycosyl transferases and carbohydrate molecules, located at the external

TABLE XIX. Activities of surface and total glycosyl transferases in gametes and vegetative cells of *Chlamydomonas moewusii* (from McLean and Bosmann, 1975)[a]

Activity	Gametes			Vegetative cells		
	Plus strain	Minus strain	0.5 vol. plus + 0.5 vol. minus	Plus strain	Minus strain	0.5 vol. plus + 0.5 vol. minus
Surface activity						
UDP-galactose	1390	1348	3378	680	720	760
UDP-glucose	1260	1294	2753	740	860	890
UDP-N-acetylglucosamine	670	1381	2806	880	260	810
CMP-N-acetylneuraminic acid	610	890	2280	600	720	700
GDP-mannose	870	1305	2319	710	720	810
GDP-fucose	1000	1140	1170	460	480	400
Total activity						
UDP-galactose	1590	1460	1346	800	840	840
UDP-glucose	1660	1392	1329	760	210	560
UDP-N-acetylglucosamine	890	1408	1290	820	860	840
CMP-N-acetylneuraminic acid	890	1104	1210	700	810	760
GDP-mannose	1140	1385	1316	810	840	840
GDP-fucose	1260	1204	1208	520	530	590

[a] Data are c.p.m./mg protein. The assay system contained 10 μl of the ^{14}C-labelled sugar nucleotide (about 10^{-9} mol), 10 μl of 0·1 M MgCl$_2$, 10 μl of 0·1 M MnCl$_2$, 10 μl of 0·1 M tris.HCl (pH 7·6) and 50 μl of active gametes or vegetative cells (about 1 mg of protein), suspended in 0·145 M NaCl. For the total system, 50 μl of a 0·1% Triton X-100 homogenate replaced the cell suspension. After 30 min at 37° C bound radioactivity was determined by adding 3 vol. of 1% phosphotungstic acid in 0·5 M HCl and counting the radioactivity in the resulting precipitate.

surface of the cell, might be responsible for cell recognition and adhesion by the formation of intercellular enzyme–substrate complexes. This could provide the high degree of specificity observed in phenomena as the one described above. Additionally, on completion of the reaction the enzyme–substrate complex would dissociate, thus explaining the temporary character of sexual adhesion and many other cell interactions.

McLean and Bosmann (1975) have tried to test this model by investigating whether glycosyl transferases are present at the external cell surface of the gametes of *Chlamydomonas*. Also, the membrane vesicles with isoagglutinating activity, budded off from the flagellar surface, were taken into consideration. Table XIX shows some of the glycosyl transferase activities present at the surface of gametes and vegetative cells, and also total activity assayed after treating the cells with Triton X-100. Total activities of individual transferases were usually about the same as the activities found at the surface, suggesting that most of the activity was located at the external membrane surface. In a mixture of *plus* and *minus* gametes most of the glycosyl transferase activities appeared to be much higher than in suspensions of single cells or in mixtures of vegetative cells. This indicates that interaction between *plus* and *minus* gametes led to an increase of carbohydrate synthesis at the cellular surface, at least in the presence of added sugar nucleotides. This phenomenon was only displayed by intact gametes, suggesting a correlation with the surface properties of these cells. The same results were obtained with mixtures of membrane vesicles isolated from gamete suspensions. However, cellular adhesion as such was not correlated with the incorporation of sugar residues in this study. As the authors point out, the presence of surface transferase activities in gamete mixtures of *Chlamydomonas* could be coincidental and be the result of heavy carbohydrate turnover and not be related to the adhesive properties of the flagellar surface. The synthetic function of these enzymes could be in the cell interior for catalysis of complex carbohydrate biosynthesis in the Golgi apparatus. Because the flagellar membrane is probably derived from the Golgi apparatus, the glycosyl transferases could thus be incidentally present at the cell surface and display activity only after the addition of sugar donors. As discussed in the preceding section, the flagellar membrane of gametes is probably subject to considerable turnover, which might explain the high transferase activities found in gametes as compared to vegetative cells.

THE FUNCTION OF FLAGELLAR ADHESION

It is attractive to suppose that the highly specific flagellar adhesion is a device evolved in this type of motile isogamous algae to promote cellular contact between compatible gametes, quite analogous to mass agglutination in ascosporogenous yeasts. If this were true, it should be possible to bypass this prelude to conjugation by sedimentation (or a similar technique, e.g. Campbell (1973)) of cells in which the adhesion has been made inoperative. In fact, Wiese and Jones (1963) have found that after treatment with EDTA (5×10^{-5} M) the formation of *vis-à-vis* pairs could be observed without prior agglutination. However, the situation is more complicated than in yeasts, because only in a very restricted area of the cell surface can cell fusion occur. From the work of Brown *et al.* (1968) it seems clear that the flagella play an indispensable role in adjusting the position of two compatible gametes to establish contact between the appropriate surface areas. Nevertheless, it is not excluded that flagellar contact between two gametes triggers some process in the cell which makes fusion possible, like the elongation of the plasma papilla in *Chlamydomonas moewusii* and *C. eugametos*. More research is clearly needed to throw light on this intriguing problem.

CHEMOTAXIS IN *CHLAMYDOMONAS*

Tsubo (1957) isolated a strain, *Chlamydomonas moewusii* var. *rotunda,* which exhibited chemotaxis. *Plus* gametes were attracted to the filtrate of a *minus* gamete suspension. *Minus* gametes, however, were not attracted. When an open glass capillary, filled with a suspension of *plus* gametes, was placed in a *minus* gamete suspension, the gametes would swim out of the capillary and clump with *minus* gametes. When *minus* gametes were placed inside the capillary tube, the *plus* gametes entered the tube from outside, and clustered there.

A comparison with other *Chlamydomonas* species, *C. moewusii, C. eugametos, C. reinhardi,* showed that interspecific chemotactic reactions were occurring between the *minus* mating type of the variety *rotunda* with *plus* gametes of other species, except *C. reinhardi.* Vegetative cells were not responsive. The attractant appeared to be volatile and could, like the attractants of *Fucus serratus* and *Ectocarpus siliculosus* (see

Chapter 9), be extracted with a stream of air. Since the attractive action could be stimulated with coal gas, the responsible factor might be a light hydrocarbon (Tsubo, 1961). No investigations were performed, however, to ascertain these observations.

7. Sex Hormones in *Volvox*

The colonial green flagellates of the family Volvocaceae are interesting material for the study of communal association of cells. Individual cells are grouped together to colonies ("spheroids" or "coenobia") which operate as well-organized units. Each colony consists of only two types of cells, vegetative and reproductive. Not only for its simplicity *Volvox* is an attractive system. It also exhibits a clearly defined sexual reproduction which can be manipulated at will in the laboratory. Sexual differentiation in these organisms is controlled by species-specific diffusible substances, the nature and action of which have successfully been investigated by Starr and collaborators.

Several species of the Volvocaceae have been under investigation, which differ in developmental respect. In this chapter particular attention will be paid to two more or less representative species, namely *Volvox aureus,* which is dioecious and homothallic, and *Volvox carteri,* which is dioecious and heterothallic. It may be recalled that in a dioecious organism only one type of sex organs is elaborated in each spheroid (male or female). In a homothallic strain both male and female spheroids are found within one clone, while in a heterothallic strain only male or female individuals are found within one clone (cf. page 4).

THE ASEXUAL SPHEROID

The cells of *Volvox* species are arranged to form hollow spheroids. *Volvox aureus,* for instance, consists of 500–2000 biflagellate cells, arranged as a single layer on the periphery of a sphere (Fig. 30). In some species, the cells are interconnected by protoplasmic strands which lie in a single

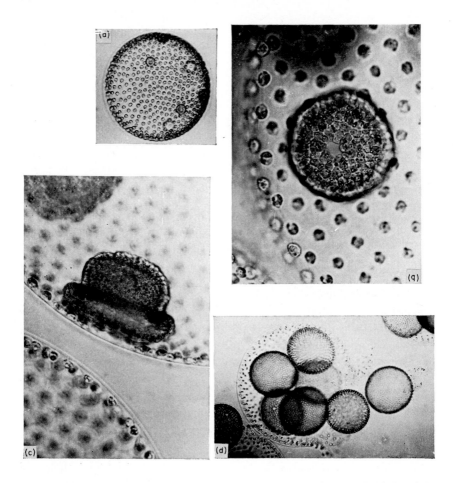

FIG. 30. (a) Freshly released daughter colony of *Volvox aureus* with undivided gonidia (×220). (b) Daughter colony in which cell division has been completed (×320). (c) Stage in the inversion process of a young daughter (×300). (d) Release of daughters by rupture of the parental colony (×65) (from Darden, 1966).

plane (Sessoms and Huskey, 1973). Two kinds of cells can be distinguished, the small vegetative cells and the larger asexually reproductive cells, the *gonidia*. The latter rarely exceed a dozen in number per colony.

The individual cells lack a normal cell wall. Rather, the protoplasts are embedded in a complex glycoprotein matrix, which keeps the cells together. The size of a colony is largely dependent on the size of the matrix. It expands considerably between the time a colony is released from the parent and its death without increase of cell number. By proteolytic enzymes a colony can be completely dissociated to single cells. Vegetative cells do not survive this treatment, whereas gonidia develop normally if care is taken to remove the enzyme by washing (Kochert, 1968).

The gonidia are the origin of daughter colonies. A series of synchronous divisions results in the formation of a hollow sphere of cells with an opening at one end, in the interior of the parent (Fig. 30a, b). The anterior ends of the cells on which the flagella will be borne are oriented inward. They become oriented outward by inversion of the whole colony (Fig. 30c). This process has been described in detail by Darden (1966, 1973) and Starr (1970). The inverted daughters continue to enlarge until they completely fill the interior of the parental colony. They are liberated by degradation of the parent (Fig. 30d).

THE MALE SPHEROID

Young male differentiated colonies can generally be easily distinguished from vegetative individuals. In male colonies of *Volvox aureus* the gonidia are completely absent. Instead, a large proportion of the cells differentiates into packets containing biflagellate sperm cells. The mature packets are released by degeneration of the parental matrix and swarm about for some time until they come into contact with a female differentiated colony.

THE FEMALE SPHEROID

In female differentiated colonies of *Volvox carteri* gonidia are replaced by egg cells, which are more numerous, smaller and appear more dense, but they may revert to the asexual gonidial state if not fertilized (Starr, 1970). In *V. aureus,* however, no differentiated female spheroids are formed. Rather, young, vegetative colonies with undivided gonidia function as such. This is evident from the fact that sperm packets are strongly attracted by vegetative colonies of the same clone (Darden, 1966). The possibility that there is a female differentiated state which in morphological respect would not be distinguished from vegetative spheroids but could be characterized by physiological properties, has not been envisaged.

Fertilization is accomplished by sperm packets penetrating the matrix of the female colonies. The packets dissociate and the individual sperm cells enter the colony where they move about actively and cluster around the egg cells, which are ultimately fertilized. Fertilized egg cells develop into zygotes which are orange coloured, and have a thick wall. They are released by degeneration of the colonial envelope.

INITIATION OF SEXUAL DIFFERENTIATION

Two observations led to the suggestion that male spheroids secrete a substance which induces (or stimulates) male differentiation in vegetative colonies. The first one was the apparent synchrony of sexual differentiation in a culture of *Volvox aureus*. Initially, a few individuals started to produce sperm packets, followed by approximately 50% of the present colonies which differentiated simultaneously. The second observation concerned transfers of differentiated cultures of this species to fresh medium. When sexually differentiated cultures were used as inoculum, very little vegetative growth occurred before colonies with sperm packets appeared in these cultures. However, when undifferentiated colonies were used, no sexually developed colonies were observed in the offspring (Darden, 1966; Starr, 1970). These observations led to the suggestion that males might trigger or stimulate male differentiation in neighbouring vegetative colonies. If a sperm-producing individual could secrete a substance committing other individuals to sexual development, the high degree of synchrony of sexual differentiation within a population would be explained. The reduction of the period of vegetative growth observed when using sexual inoculum could be caused by a carry-over of this substance.

Similar observations were made in other species. In the heterothallic species *Volvox carteri* and *rousseletii*, for instance, also a more or less synchronous female differentiation, together with male development in a mixed natural population, was reported (Starr, 1970; McCracken and Starr, 1970). Again, this happened after spontaneous appearance of a few sexual males, which suggests that these individuals influenced in some way sexual development in other individuals. With one exception (*V. dissipatrix*; Starr, 1972a), sexually developed females were not observed to stimulate sexual differentiation in neighbouring individuals.

THE SEX FACTOR OF *VOLVOX AUREUS*

In 1966, Darden demonstrated that sexually differentiated cultures of *Volvox aureus* contain a substance in the medium which stimulates male development in this species. One ml of cell-free medium was inoculated

TABLE XX. Male differentiation of *Volvox aureus* in medium derived from sexually developed cultures (from Darden, 1966, modified)

Experiment	Percentage of male differentiated spheroids	
	Fresh medium	Medium from sexually developed cultures
1	0	45
2	0	47
3	0	45
4	0	45
5	0	57
6	0	33
7	0	58
8	0	24
9	0	44
10	0	49
11	0	36
12	0	41
Average	0	44
Total colonies	3320	3383

with five vegetative colonies of comparable age. These were obtained by rupturing parental colonies containing mature daughters by forcing them through a drawn-out pipette. The released daughters served as inoculum. At the end of one generation (approx. 96 h), each sample was scored for male colonies. The results of some experiments of this type are presented in Table XX. They show that inocula incubated in the filtrate of a sexually developed culture gave rise to populations containing an average of 44% males, whereas in the controls (incubated in fresh medium) no single male could be detected.

The possibility that the incubation in depleted culture medium had led to sexual differentiation became less probable by the subsequent isolation of a substance inducing male differentiation (Darden, 1966; Ely and Darden, 1972). This was made possible by the development of an assay procedure which was based on the above-described experiment. Serial dilutions of test material with fresh medium were inoculated with ten vegetative colonies of *Volvox aureus* of comparable age. One generation period later, the progeny was scored for sexual males. Since 50%

TABLE XXI. Concentration and purification of the sex factor from *Volvox aureus* (from Ely and Darden, 1972)

Solution	Titre[a]	Carbohydrate (μg/ml)	Protein (μg/ml)	Specific activity $\times 10^5$
Culture medium	$2\cdot6\times10^6$	11·5	13·0	2·0
Concentrated medium	$7\cdot9\times10^8$	290·0	290·0	27·2
Sephadex G200 peak	$4\cdot3\times10^7$	35·6	29·5	14·6
DEAE Sephadex peak	$3\cdot0\times10^7$	15·0	12·5	24·0
SE Sephadex peak	$3\cdot0\times10^7$	12·4	8·3	36·1

[a] Defined as that dilution which results in the production of 25% male colonies.

was the maximal value to be obtained, the unit concentration of male-inducing substance was defined as that concentration which brought about sexual development in 25% of the present colonies.

Why only part of a clone could be brought to produce sperm packets in *Volvox aureus* is unknown. As mentioned above, the non-differentiating individuals serve as females. In heterothallic species (in which all members of a clone are male *or* female), like *V. carteri* (Starr, 1969) or *V. rousseletti* (McCracken and Starr, 1970), virtually 100% induction is realized in the male and female strains with the sex factor of these species (see below).

The sex factor of *Volvox aureus* was purified as follows. Medium from sexually developed cultures was concentrated by freeze-drying and dialysed. Purification was accomplished by chromatography on Sephadex G200 and the ion exchangers DEAE- and SE-Sephadex. From all columns a single band of biologically active material was eluted, with high molecular weight, and with cation as well as anion properties. The most purified product contained protein and polysaccharide, suggesting a glycoprotein nature of the active species. Biological activity was destroyed by the action of pronase (Ely and Darden, 1972). In Table XXI the effect of concentration and purification of the active substance is demonstrated.

SENSITIVITY OF *VOLVOX AUREUS* TO ITS SEXUAL FACTOR

It appeared that the sensitivity of vegetative spheroids of *Volvox aureus* to the sex factor isolated from sexually developed cultures is a function of their life cycle. Darden (1966) demonstrated that an optimum in susceptibility exists in this species by incubating individuals at various developmental stages. Freshly released colonies were allowed to develop

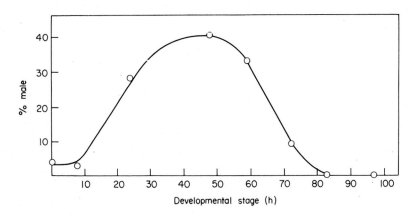

Fig. 31. Susceptibility of a synchronous vegetative culture of *Volvox aureus* towards a 30 min exposure of the sex factor (from Darden, 1966).

in fresh culture medium. Samples were taken at various times and incubated for 30 min in a medium containing the sex factor, after which they were washed and reincubated in their original medium. At the end of one generation period all samples were scored for percentage of males. The results are presented in Fig. 31. They indicate not only that a 30 min exposure was sufficient for induction of male differentiation, but also that there was a remarkable difference in sensitivity to the action of the sex factor. Microscopical observations suggested that maximum susceptibility coincided with a maximum rate of cell division. Since the major morphological difference between vegetative and male colonies is the formation of either gonidia or sperm packets during the period of cell division, Darden (1966) suggested that the optimum in the sensitivity curve might coincide with this branch point in development.

THE SEX FACTOR OF *VOLVOX CARTERI*

Contrary to the species discussed above, in *Volvox carteri* both male and female differentiation is stimulated by a substance secreted by male colonies. The availability of strains that are purely male or female made it possible to prove that the factor involved in sexual differentiation was only produced by sexual males (Starr, 1969). It appeared to be secreted during sperm packet release (van de Berg and Starr, 1971). It was assayed by means of the female strain, which shows no spontaneous sexual differentiation, as male strains do. Serial dilutions of fluid to be tested were made and inoculated with 75 freshly released spheroids. Using specified conditions, described in detail by Starr (1969) and Starr and Jaenicke (1974), this inoculum gave a total of approx. 1000 spheroids in two generations. These were scored for sexual development (formation of eggs instead of gonidia). When applying non-limiting amounts of the sex factor, 100% induction was obtained.

The production of the active substance from the variety *nagariensis* of *Volvox carteri* by Starr (1969) routinely involved the transfer of asexual male spheroids from small to larger volumes, and the addition of male fluid from a previous sexual population to a homogeneous population containing individuals with young spheroids just ready to be released. Male differentiated spheroids formed within 48 h, approaching 100% of the total population. The release of male spheroids was followed within a few hours by the release of sperm packets, which resulted in a suspension of sperm packets and vegetative cells. The fluid was harvested 36 h after total release of the sperm packets. With a special isolate (Starr and Jaenicke, 1974), characterized by high sex factor production, crude culture media were obtained which gave 50% induction at a 10^{-9}-fold dilution in the assay procedure described above.

The active substance was concentrated by passing the culture fluid through a column of carboxymethyl cellulose, equilibrated at pH 5. All of the activity remained on the column. It was eluted with 0·1 M NaCl solution and was further purified by Sephadex G50 and G75 gel filtration. The purified material contained approx. 55% protein and 45% carbohydrate. The molecular weight, by comparison with standard proteins on Sephadex G25 was estimated to be 25 000. Polyacrylamide gel electrophoresis in SDS gave a well-defined single band. From its mobility a molecular weight of 28 400 was derived. Sucrose gradient centrifugation gave a molecular weight of 26 000, assuming a partial specific volume of 0·7. On sedimentation velocity analysis the highly purified preparations formed two symmetrical bands with $s^0_{20,w} = 1·57$ and $2·7$ S, respectively. Both of them were biologically active. For the slow band a molecular weight of 30 600 was calculated, the other band probably being a dimer. The amino acid analysis did not reveal unusual percentages of amino acids. Gas chromatograpic carbohydrate analysis of the purified material gave as the major sugars xylose (25%), mannose (16%) and glucose (32%).

The maximal biological activity recorded of a purified preparation was

14·4% induction at a concentration of 10^{-11}g/litre. In Starr's experimental procedure, involving 1000 colonies per 10 ml of assay fluid, this means that approx. 18×10^5 molecules gave 14·4% response in 1000 individuals (Starr and Jaenicke, 1974).

The factor isolated from the variety *nagariensis* appeared to exhibit considerable specificity even among different isolates of the same species. This suggests that inducing factors of various isolates may be different (Starr, 1970). Incidently, Kochert and Yates (1974) purified the sexual factor of another strain of *Volvox carteri*, but no major differences were found. They used essentially the same purification procedure as Starr and Jaenicke. Their preparations appeared to be homogeneous as tested by polyacrylamide gel electrophoresis in urea. By electrophoresis in SDS the molecular weight was estimated to be 32 000, which was confirmed by sedimentation equilibrium centrifugation.

The availability of the sexual factor of *Volvox* species in highly purified form widens the opportunities to study the biochemical background of its action. Apart from immunological methods which can now be used, it appears that the factor can be labelled with iodine without loss of biological activity (Starr and Jaenicke, 1974).

CONTROL OF SEXUAL DIFFERENTIATION IN *VOLVOX*

As in *Volvox aureus,* sexual development in *V. carteri* starts with the spontaneous appearance of a few sexual males, which in turn induce male and female differentiation in other individuals. How sexual differentiation is induced in these primary males is unknown. It is possible that environmental factors are involved. Van de Berg and Starr (1971) report, for instance, that the male strain of *V. gigas,* when grown on a defined medium, must be transferred frequently to keep it in an asexual condition. Rarely can more than two generations be grown in a culture flask without the entire population becoming sexual. When grown in soil–water medium, however, male spheroids are seldom encountered. This suggests that nutritional factors exert a strong influence on sex expression. Starr (1972b), in contrast, has suggested that in *V. carteri* var. *nagariensis* spontaneous changes at the gene level might give rise to the appearance of the initial sexual males. He observed also that in female strains occasionally spontaneous sexual differentiated spheroids are produced, at a frequency of less than 1:10 000. The appearance of such spontaneous females could be attributed to environmental factors, or be caused by mutation.

In the latter case it should be possible to establish clonal populations of the female strain which would permanently bear the female sexual characteristics. Starr demonstrated that some eggs of such spontaneous females did revert to the asexual (i.e. gonidial) state when transferred to fresh medium, as normal unfertilized females do; however, such eggs produced new individuals which were not asexual spheroids with gonidia but typical female spheroids with eggs. Successive generations in these strains produced only females, suggesting that a mutation had occurred and that it could be inherited through asexual reproduction. In one case, the mutation appeared not to be sex-linked so that the mutant locus could be passed through the male and then back to the female.

Whether a similar genetic change is the basis for the appearance of sexual males remains to be seen, but is not considered improbable by Starr, in view of the fact that the males react to the same substance and at the same levels of concentrations as females do. This would imply that sexual differentiation in this strain of *Volvox* is initiated by a spontaneous change in the genome rather than by some (combination of) environmental conditions.

As in many other hormonal systems, the sex factor of *Volvox* interrupts the vegetative growth cycle and causes completely different potencies of the cell to be expressed. In female colonies of several species gonidia are inhibited by the factor to divide and are forced to follow another developmental route, leading to gametes. Also in male colonies, gonidia exhibit a completely different development in the presence of the factor, resulting in sperm packets which are only superficially homologous to the embryos which are the result of undisturbed vegetative development. Thus this system provides excellent opportunities for further study of the molecular basis of development in a primitive organism (Starr, 1972b; Darden, 1973).

8. Sexual Behaviour of the Green Alga *Oedogonium*

Oedogonium is a most remarkable genus as regards the interaction between male and female sexual cells, and it has been treated in several texts (e.g. Wiese, 1969), without much evidence being available what exactly the nature is of this interaction. The work that has been done on this group of algae is so admirable that it will be mentioned briefly.

Oedogonium species are freshwater green algae, the thallus consisting of long unbranched filaments. Vegetative development occurs by means of fragmentation and production of multiflagellate zoospores. Sexual reproduction proceeds by oogamy. In monoecious forms the oogonia and antheridia are produced on the same plant, while in dioecious forms the oogonia and antheridia appear on different filaments which are morphologically alike. In *macrandrous* species, the antheridia produce motile sperm cells which are chemotactically attracted by the oogonia. In

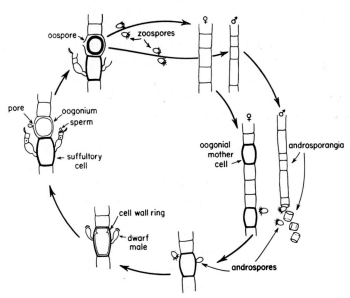

Fig. 32. Sexual reproduction in *Oedogonium borisianum* (from Rawitscher-Kunkel and Machlis, 1962)

E

nannandrous species, however, the situation is more complex. Male-determined cells form so-called androspores, which are similar but smaller than zoospores. These androspores settle on the wall on or near to an oogonium mother cell, and germinate into a small male plant, consisting of one or two cells. This dwarf plant gives rise to a few motile sperm cells from an antheridium, which are trapped by the gelatinous sheath surrounding the oogonium. This type of sexual behaviour is illustrated in Fig. 32. Details about the ultrastructure of fertilization in *Oedogonium* and related species have been reported by Hoffman (1974) and Retallack and Butler (1973).

In nannandrous and macrandrous species sexual differentiation is initiated, irrespective of the presence of the complementary sex, and is induced by environmental factors. *Oedogonium cardiacum*, for instance, must be resuspended in a growth medium lacking nitrate and illuminated to start sperm and oogonia production. In addition, contamination of the cultures with certain bacteria seems to be a necessary condition for reproductive organs to be formed (Machlis, 1973c).

Chemotaxis of sperm cells in the macrandrous *Oedogonium cardiacum* was first studied by Hoffman (1960). When a capillary tube was filled with water, in which female plants had produced oogonia, and was placed in a suspension of sperm cells, these accumulated at the tips and eventually entered the capillary. Machlis and coworkers (1974) described some experiments with the attracting agent. They estimated its concentration in a semiquantitative manner by exposing agar plugs, containing the attractive substance, to a sperm suspension. In a very strong reaction virtually all the sperm cells accumulated in the immediate vicinity of the attractant. The substance appeared to be highly water-soluble, non-extractable by common organic solvents, and to be rapidly destroyed at high temperatures or extreme pHs. Its molecular weight was estimated to be between 500 and 1500.

In the nannandrous *Oedogonium borisianum* at least four stages of sexual interaction are involved, according to Rawitscher-Kunkel and Machlis (1962). (1) The attraction of the androspores must be a response to a diffusible substance emanating from the oogonial mother cells. This can be demonstrated by filling a capillary tube with liquid from a fertile female culture and immersing it in a suspension of androspores. (2) After a period of active swarming, the androspores attach to the cell wall at or near an oogonial mother cell. Shortly later, the androspores are seen to elongate and to produce an antheridium. The direction of elongation seems to be dependent on the position of the spore with respect to the oogonial mother cell. Androspore development proper is completely independent of the influence of oogonial mother cells. Androspores kept in the same male culture from which they originated can develop into dwarf males and produce sperm. It is also possible to obtain normal-sized vegetative male filaments by transferring androspores to fresh enriched nutrient medium. (3) Oogonial development, however, is not independent of the male's presence. An oogonial mother cell only forms an oogonium if androspores are attached. When a female filament with androspore-attracting mother cells is covered with a thin layer of agar and placed in a suspension of androspores, these are attracted but do not come into contact with the oogonial mother cell. No development into oogonia is then observed. If the agar surrounding the filament is opened with a glass needle, andospores attach to the oogonial mother cell which consequently divides within the next 24 h. (4) The newly formed oogonium is surrounded by a mass of mucilage in which the sperm cells, produced after development of the dwarf males, are trapped. Primarily, the sperm cells exhibit a random motion but after a while they aggregate close to the oogonial pore. From these cells one is caught by a protoplasmic papilla appearing through the pore. This suggests that the sperm cells are attracted to the pore by some factor released by the oogonium.

9. Chemotactic Hormones in Some Brown Algae

Fucus serratus is a well known seaweed found on the shores of the Northern Hemisphere. The plant consists of a long, dichotomously branched thallus attached to a solid surface by means of a basal disc. At the tips of the thallus fruit bodies are formed, called conceptacles, which contain several oogonia or antheridia. *Fucus serratus* is dioecious, the two types of sex organs occur on separate plants. The expulsion of eggs and sperm cells takes place at low tide. Because of loss of water the thallus shrinks so that the oogonia and antheridia are squeezed to the surface, together with a mass of mucilage, from the inner part of the conceptacles. When the tide returns both egg and sperm cells are liberated from their envelopes into the water.

The sperm cells have a long propelling flagellum and also a short one which might serve as a steering device. They are chemotactically attracted to the egg cells and adhere to them by means of their long flagellum prior to fertilization. This indicates that the egg cells secrete a diffusible compound which attracts sperm cells. Cook and Elvidge (1951) demonstrated the existence of such an attractive compound, and also showed that it was quite volatile. The sperm-attracting action was mimicked by several hydrocarbons such as *n*-hexane (in a dilution of 10^{-6}–10^{-7}M) and also some ethers and esters. But none of these substances appeared to be fully identical with the natural factor. Also Müller (1972) demonstrated the hydrophobic and volatile character of the attractant by showing that it could be extracted by a stream of air from an aqueous solution. When the air was subsequently passed over Vaseline droplets, the attractant was absorbed, as was shown by placing the droplets in a suspension of sperm cells. The droplets acted like eggs and became surrounded by a halo of sperm. This procedure provided a means to assay the attractant in a qualitative fashion.

Müller and Jaenicke (1973) were able to isolate the attractant in measurable quantities and to elucidate its structure. Female plants were collected at the Marine Station of Roscoff (France), the fruiting tips were isolated and soaked in sea water until the eggs and oogonia were released. No attempt was made to separate the eggs from the oogonia, since also oogonia with unreleased eggs attracted sperm cells. Through the resulting suspension a stream of air was drawn. The air stream was subsequently led through a series of three cold traps in which the volatiles from the egg suspension were condensed. In the first two traps water vapour and some attractant was condensed, while the third one, containing an inert liquid fluorinated hydrocarbon, efficiently washed the attractant from the air stream. The solution thus obtained was subjected to preparative gas chromatography, which yielded one homogeneous band of the active substance. From 252 kg of fresh female tips, 9·8 litres of eggs were derived, which gave 690 μg of attractant. By mass spectrometry and chemical analysis the compound was identified as 1,3,5-octatriene. A synthetic sample proved to be identical with the natural product in its chemical and biological properties. In a subsequent study, Jaenicke and Seferiadis (1975) synthesized all four stereoisomers of this compound and compared them with the natural compound by means of analytical gas chromatography. They thus were able to prove that the structure of the hormone is 1,3-*trans*, 5-*cis*-octatriene:

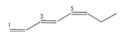

Structure of fucoserratene

The compound was named *fucoserratene* by the authors.

The brown alga next studied, *Ectocarpus siliculosus*, consists of branched uniseriate filaments, like *Fucus* widely distributed along coasts. It is also known as a major ship-fouling species. Diploid plants contain sporangia which after meiosis give rise to numerous haploid motile cells which may either function as gametes, or else develop without fertilization to a haploid plant. Diploid plants also produce diploid zoospores which germinate directly. Haploid plants contain gametangia which give rise to isogametes. The species is strictly dioecious, male and female sex organs occurring on separate plants (Müller, 1967).

Initially, isogametes of both sexes are equally motile and do not show any mutual attraction. However, female gametes eventually become sessile and attach to a solid surface. From then on, they strongly attract male gametes. These cluster around the female cells in large numbers and adhere to them with the front flagellum. As soon as one male gamete has penetrated a female cell, the others become detached and evidently are no longer attracted.

Müller (1968) demonstrated that female gametes secrete an attractant which, like fucoserratene, could be isolated by passing air through a gamete suspension. Biological assay was performed by drawing the air along a droplet with male gametes, which in response formed transient clumps, preceded by vehement swimming movements in circles near the surface of the droplet. Another way of assaying the attractive compound was by exposing one end of an open capillary tube, filled with a male gamete suspension to attractant-containing air. The male gametes concentrated at this end of the capillary, again exhibiting circular movements and temporary clumping. These reactions have not been investigated in depth.

It is interesting to note that female gametes evidently start to secrete the attractant only after having become immotile. What exactly triggers this secretion is unknown. According to Hartmann (1956), the behaviour of female gametes to become sessile is only observed when both types of gametes are mixed, which suggests that male gametes have some inductive action on female gametes. The secretion of attractant probably is arrested immediately after fusion, because chemotaxis then stops abruptly. The phenomenon of the sudden withdrawal of the attracted gametes after fertilization could also be caused by the emission of a substance causing negative chemotaxis (Levring, 1952). It is also possible that a change in the surface of the female gamete occurs after fertilization, causing detachment of the excess male cells. Wiese (1969) has suggested that the attracting substance might be involved in the adhesion of male and female gametes (cf. Chapter 1).

The chemotactic substance of *Ectocarpus siliculosus* was isolated as follows. A female gametophytic clone was grown in large quantities in culture dishes. The attractant was removed from the cultures by means of a stream of purified air and condensed in a cold trap at $-80°$. The condensate was then flushed with a stream of nitrogen through a drying tube containing anhydrous $CaCl_2$ and dissolved in carbon tetrachloride. The active compound was purified by preparative gas chromatography. With this method, 92 mg of material was collected. By means of mass spectrometry and nuclear magnetic resonance spectroscopy the compound was identified as all-*cis*-1-(cycloheptadiene-2',5'-yl)buetene-1 (Jaenicke *et al.*, 1971, 1973):

The compound still showed activity at a concentration of 10^{-12} M. A racemic synthetic preparation appeared to be biologically active and to exhibit the same chromatographic and chemical characteristics as the natural product (Müller and Jaenicke, 1973). The natural substance showed optical activity, however $(\alpha_D^{22} = +72°)$. The compound was renamed *ectocarpene* by Müller and Jaenicke (1973).

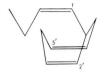

Structure of ectocarpene

The biological action of ectocarpene is not very specific. Müller (1968) showed that a large variety of hydrocarbons attract male gametes in low quantities (which still are many orders of magnitude higher than the minimal concentration of ectocarpene itself). Several alkylcycloheptadiene analogues of ectocarpene also showed some activity. Interestingly, however, analogues with long side chains showed an inhibitory action and might be competitive inhibitors, binding to the same reaction site as ectocarpene (Jaenicke, 1975).

By exposing male gametes of *Ectocarpus* for different lengths of time to a gas phase containing ectocarpene, labelled with tritium in the side chain, the kinetics of uptake of the hormone was studied (Jaenicke, 1974). Short exposure times gave high specific labelling, expressed as d.p.m. per cell. With longer exposure times (more than about 15 s), the labelling diminished, apparently by some metabolic degradation of the side chain, until a low steady state was reached. The time course corresponded more or less to the responsiveness of the gametes as a function of time. This seems strongly suggestive of a specific messenger–receptor mechanism in which the messenger is rapidly metabolized. The reader is reminded of an analogous situation in the chemotactical system of *Allomyces* (Chapter 2). According to preliminary autoradiographic evidence, most of the rapidly accumulated radioactivity was located in the long flagellum, which is used by the male gametes for propagation and adhesion to the female cell (Jaenicke, 1974).

A third chemotactical hormone was identified by Jaenicke and co-workers (1974) from *Cutleria multifida.* In this anisogamous and dioecious species the female gametes, after liberation from the gametangia, remain motile for a period which can vary between 5 min to 2 h, while the male gametes can remain active for about 20 h. No fertilization occurs while the female gametes are motile (Chapman and Chapman, 1973). After settling, they round off and attract large numbers of sperm cells chemotactically for 3–4 h. From this plant large amounts of egg cells were obtained by cultivation in the laboratory. By the same method applied so successfully to isolate fucoserratene and ectocarpene, three compounds were isolated from a suspension of *Cutleria* eggs, one of which being biologically active. From mass and proton magnetic resonance spectrometry this compound was identified as *trans*-4-vinyl-5-(*cis*-1′-butenyl)-cyclopentene. As a trivial name *multifidene* was proposed by the authors.

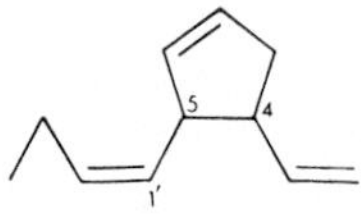

Structure of multifidene

The two other constituents of air passed through a suspension of egg cells appeared to be structurally closely related. One of them exhibited not only the same mass spectrum (which was virtually identical for all compounds) but also had an identical gas chromatographic retention time and proton magnetic resonance spectrum as ectocarpene. The second inactive compound, named *aucantene,* was identified as *trans*-4-vinyl-5-(*trans*-1′-propenyl)cyclohexene:

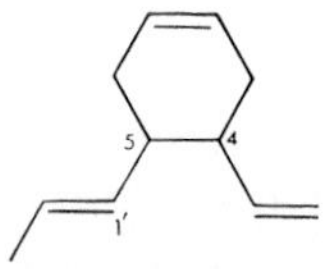

Structure of aucantene

Both inactive compounds were also found to be constituents of asexual plants of *Cutleria multifida*.

The close similarity of the metabolites isolated from various brown algae not only suggests a common biogenetic origin (cf. Jaenicke *et al.,* 1974), but also a strong specificity of action. However, little experimental material is available about this matter. On the other hand, it is clear that, as Jaenicke (1974) points out, these simple, albeit quite subtle systems may be useful in the elucidation of chemotaxis, chemical transmission and, as models, for other sensory processes.

10. Sex Hormones in Ferns: Antheridiogens

As stated by Miller (1968) in a review article, fern gametophytes are admirably suitable as experimental objects for the study of plant growth and differentiation. They are morphologically very simple, consisting of only a few types of cells, and they can be handled in large numbers in sterile culture. This makes it possible to perform experiments in accurately controlled culture conditions.

The haploid, gametophytic generation of ferns begins when a spore germinates and forms a protonemal cell and a primary rhizoid. The protonema forms in most species a single-layer thallus, the prothallus, with uncellular rhizoids at the lower side. On this body a multitude of antheridia and oogonia originate (Fig. 33). Fertilization takes place with motile sperm which are chemotactically attracted to the egg cells. From the fertilized egg the diploid sporophyte develops. Methods to cultivate prothalli on a wide variety of solid and liquid media have been reviewed by Miller (1968).

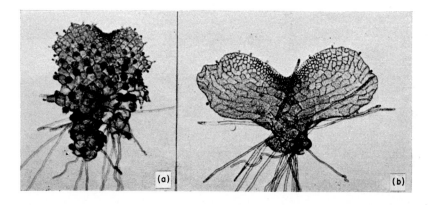

Fig. 33. Gametophyte of *Onoclea sensibilis*, 19 days old ($\times 33$): (a) grown in the presence of *Pteridium* medium (1:250); (b) grown in the absence of *Pteridium* medium (from Näf, 1956).

Although prothalli are in principle able to produce both antheridia and oogonia, the proportions in which these sex organs arise is dependent on many factors. Frequently there is a correlation between poor growth rate and relative abundance of male sexual structures. Also, conditions unfavourable for rapid growth often lead to exclusively male plants. However, when the plants are returned to more favourable conditions, they may develop both type of organs.

In *Polypodium crassifolium,* antheridium formation is influenced by the light quality to which the prothallus is exposed. This light control seems to be phytochrome-mediated, since far-red light, or darkness, which promote antheridia formation, are counteracted by red light (Schraudolf, 1967).

Archegonia develop usually later than antheridia, so that at some late stages of development prothalli may seem to be unisexually female (Näf, 1958).

THE ANTHERIDIOGENS

Antheridium formation in many ferns is controlled by naturally occurring substances which are called *antheridiogens* (Pringle, 1961). Döpp (1950) first demonstrated that extracts from mature prothalli of *Pteridium aquilinum* stimulate the formation of antheridia in young prothalli of the same fern. Also, the substrate on which prothalli were growing appeared to contain this activity (Döpp, 1959). It could be diluted as much as 30 000 times before the antheridia-inducing activity disappeared if assayed against a related species, *Onoclea sensibilis* (Fig. 33, Näf, 1956). This suggests that prothalli secrete a substance which hastens the onset of antheridium formation in neighbouring prothalli. The *Pteridium* extract also exhibited activity on *Dryopteris filix-mas.* While under normal conditions antheridia arise on gametophytes of this fern only after 6–8 weeks, these sex organs are produced 10–14 days after inoculation of the prothalli in the presence of this factor (Döpp, 1950). Other Polypodiaceae appear to be sensitive to this factor, although the threshold concentrations vary widely (Näf, 1968; Näf et al., 1975). *Onoclea sensibilis* is particularly interesting because this species produces no antheridia at any stage of development under specified culture conditions but does so when an inducing extract from other species is applied (Näf, 1958). For this reason, this species is frequently used to assay the active factor involved. Also, the apogamous fern *Notholaena sinuata* forms antheridia on medium supplemented with *Pteridium* extract, even though none is seen to arise spontaneously (Döpp, 1959).

The factor isolated from polypodiaceous species appeared to be inactive towards species of other families, with some exceptions (listed by Näf, 1962a; Miller, 1968). *Anemia phyllitidis*, of the Schizeaceae, on the other hand, elaborates an antheridiogen which is inactive towards members of the Polypodiaceae, including *Onoclea sensibilis* (Näf, 1959). In *Lygodium japonicum*, the formation of antheridia is controlled by still another substance, even though this species belongs to the Schizeaceae (Näf, 1960, 1968). Schedlbauer (1974) demonstrated that the antheridiogen isolated from *Ceratopteris thalicroides* (Parkeriaceae) is distinct in its biological activity from the *Pteridium* and *Anemia* antheridiogens. This suggests that antheridium formation is controlled by different substances in different groups of ferns. For that reason the terminology suggested by Näf (1968), which is analogous to that used for gibberellins, seems most appropriate: A_{Pt} for the antheridiogen of *Pteridium*, A_{An} for that of *Anemia,* etc.

According to Döpp (1959), the antheridiogen A_{Pt} hastens the formation of antheridia, but does not affect the production of archegonia, even at extremely high concentrations. Other antheridiogens do not seem to influence archegonia formation. High concentrations of antheridiogen A_{Pt}, however, inhibit vegetative growth of the gametophyte (Döpp, 1950).

BIOASSAY OF ANTHERIDIOGENS

The assay of biological activity is based on the ability of antheridiogens to induce antheridia in young prothalli. As test organism of antheridiogen A_{Pt} *Onoclea sensibilis* is preferably used because, as mentioned, it does not produce antheridia spontaneously (see, however, page 139). For the antheridiogen A_{An}, *Anemia phyllitidis* is used as assay organism. The preparation to be examined is added to culture medium in a series of dilutions, autoclaved, and inoculated with spores. After 20 days the percentage of prothalli carrying antheridia is determined (Näf, 1966). In some reports, the prothalli are exposed at the age of 9 or 10 days, which leads to a shortened lag period between exposure to the antheridiogen and the onset of antheridium formation (Näf, 1967; Endo et al., 1972).

Another assay procedure sometimes used is based on the ability of A_{An} to substitute for light requirement in spore germination. In this, as in other respects, this antheridiogen strongly resembles the gibberellins (Schraudolf, 1964; Näf, 1966). In this assay, which is some 30-fold more sensitive than the first one described, the percentage of germinated spores is determined 5–6 days after germination.

ISOLATION AND CHARACTERIZATION OF ANTHERIDIOGEN A_{An}

The isolation and structural elucidation of the antheridiogen produced by *Anemia phyllitidis* was carried out by Endo and coworkers in 1972. The raw material, from which the hormone was extracted, was obtained as follows. Erlenmeyer flasks, containing 38 ml of solidified mineral medium, were densely inoculated with *Anemia* spores. After 22 days, some of the prothalli were transferred to new culture flasks with the same amount of medium. Thirty-one days thereafter, 8 ml water was added to each flask. After 4 h at room temperature, the flasks were stored overnight in the freezer. Following thawing, the liquid was filtered off. The filtrate induced antheridia to a dilution of 1:1000. A 30-litre batch of this material was extracted with ethyl acetate and the extract concentrated *in vacuo*. The residue was subjected twice to thin-layer chromatography. The active material thus purified gave the composition of $C_{19}H_{22}O_5$ on elementary analysis. Subsequent spectroscopic and chemical studies of the isolated compound led to the following structural formula (Nakanishi *et al.*, 1971):

The antheridiogn of *Anemia phyllitidis* (A_{An})

The purified product induced antheridia to a dilution of 10 μg/litre and also substituted for light requirement in spore germination to a dilution of 0·3 μg/litre. A total of 1500 flasks as described above gave a yield of 18 mg. The related species *Anemia hirsuta* appeared to elaborate an antheridiogen with the same structure (Zanno *et al.*, 1972).

The structure of antheridiogen A_{An} is very similar to that of the gibberellins. Its chromatographic behaviour on silica gel thin layers is similar to that of gibberellin A_3. This chemical similarity is reflected by the biological activities of these compounds. Apart from the similar effect on spore germination, mentioned above, gibberellins also induce antheridia in

Anemia prothalli (Schraudolf, 1964; Voeller, 1964a). According to Voeller and Weinberg (1967), the effects of gibberellin A_3 is indistinguishable from the effects of the naturally occurring hormone: the time interval between application of and the response to both substances is quite similar, and the dosage–response curve of the antheridiogen is paralleled by that of gibberellin A_3. The threshold activities, however, are quite different: the lowest concentration of gibberellin A_3 giving a response is 1 mg/litre (Näf, 1966), while this figure of antheridiogen A_{An} is 10 μg/litre. The most active gibberellin appeared to be A_7, followed in decreasing relative activity by the gibberellins A_4, A_9, A_3, A_1, A_5 and A_8, as is shown in Table XXII (Schraudolf, 1966a).

TABLE XXII. Minimal concentration of gibberellins showing antheridia-inducing activity in prothalli of *Anemia phyllitidis* (from Schraudolf, 1966a, modified)[a]

Gibberellin	Concentration (g/ml)
A_1	5×10^{-7}
A_3	5×10^{-7}
A_4	5×10^{-9}
A_5	5×10^{-5}
A_7	5×10^{-10}
A_8	5×10^{-5}
A_9	5×10^{-7}
Allo gibberellin	$2 \cdot 5 \times 10^{-6}$

[a] Prothalli were cultivated floating on a mineral medium in continuous light. Decreasing concentrations of the compounds were added while preparing the medium. At 24 h intervals 100 prothalli were sampled at random and the ratio of prothalli with and without antheridia was determined.

Nakanishi *et al.* (1971) and MacMillan (1974) have pointed out that antheridiogens can be thought to be derived from gibberellins, in the following way:

Gibberellin A_4 Antheridiogen A_{An}

This leads to the supposition that the biological activity which is observed on gibberellin application to *Anemia* prothalli is the consequence of this or a similar conversion of gibberellin to antheridiogen.

In contrast, gibberellins have much less effect in species of other families such as the Polypodiaceae (Voeller, 1964a). The antheridiogen isolated from these species (specifically A_{Pt}) is chemically distinct from A_{An} in its chromatographic behaviour and biological properties (Voeller, 1964b). Nevertheless, it seems probable that its structure is similar to that of gibberellins. Studies by Pringle and coworkers established that this sub-stance also contains a carboxyl group as well as one double bond, and can be extracted with organic solvents (Pringle *et al.*, 1960; Pringle, 1961).

A third antheridiogen, isolated from culture media of *Onoclea sensibilis*, a species which is also sensitive towards the antheridiogen A_{Pt}, appeared to differ in its mobility on thin-layer chromatography (Näf *et al.*, 1969). Untreated medium of 10-day-old plants of this species contains very little, if any, active material. Biological activity is only detected if the medium or prothallial extract, from which the factor is derived, is autoclaved. A natural antheridium-inducing factor seems not to be involved, since young,

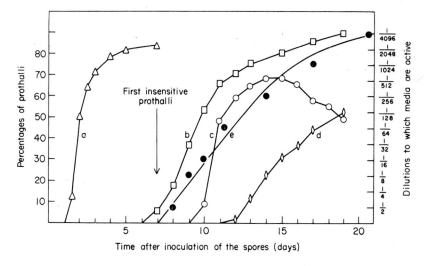

Fig. 34. Relationship between different developmental features and antheridiogen production in *Pteridium aquilinum*. a, Percentages of germinated spores; b, percentages of heart-shaped prothalli; c, percentages of prothalli bearing antheridia (or their initials); d, percentages of prothalli bearing archegonia (or their initials); e, dilutions of *Pteridium* medium which brought about the formation of antheridia in 50% of the assay prothalli (from Näf, 1958).

sensitive prothalli, growing amidst 10-day-old prothalli, do not produce antheridia. The possibility that a natural inhibitor is released by the older prothalli, obscuring an active factor, was ruled out by the finding that non-autoclaved *Onoclea* medium failed to inhibit antheridia formation when added together with autoclaved medium (Näf, 1965).

THE ROLE OF ANTHERIDIOGENS IN SEXUAL REPRODUCTION OF FERNS

Näf (1956) has studied the relationship between antheridium formation and the production of antheridiogen in prothalli of *Pteridium aquilinum*. Figure 34 illustrates some features of gametophyte development. The presence of antheridiogen is first detected two days before antheridial initials can be distinguished, and about one day after the first appearance of heart-shaped prothalli. Once started, the accumulation of antheridiogen proceeds at a rapidly increasing rate which gradually declines again after about 10 days. Seven weeks after inoculation of the spores a maximum of activity is observed.

The sensitivity, on the other hand, of *Pteridium* prothalli does not coincide with maximum production rate. Treatment of prothalli at different ages with active solutions showed that the plants become completely insensitive at a time at which they attain a heart-like form. Also the following experiment demonstrates this. One hundred randomly chosen prothalli (twenty-five out of each of four 9-day-old cultures) were transferred, one per 50 ml flask, to antheridiogen-containing medium. Prothalli isolated at such an early stage of development all failed to form antheridia spontaneously but they produced them readily in response to added antheridiogen. Examination of the plants for antheridium formation occurred 5 days following exposure to the active factor. It turned out that the prothalli which had not yet the heart-shape at the time of transfer all formed antheridia, whereas other individuals with a slightly faster developmental rate, and which were heart-like at the time of transfer, failed to form antheridia. A similar conclusion was derived for prothalli of *Onoclea sensibilis*. The relationship between the shape of prothalli and antheridia production is shown in Fig. 35.

It may be noted that the presence of antheridiogen in the medium is not detected until one day after the prothalli have become insensitive to it. This suggests that they become insensitive to the factor before they begin to produce it in detectable quantities. Also in other cases spontaneous

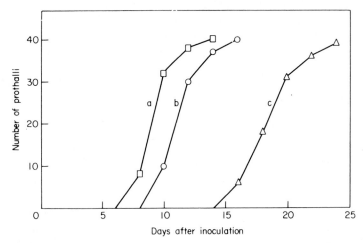

Fig. 35. Relationship between the decline of sensitivity towards antheridiogen and the attainment of heart-shape and archegonia production. a, Percentages of prothalli that have attained heart-shape; b, percentages of prothalli that have become insensitive to the antheridiogen; c, percentages of prothalli that have initiated one or more archegonia (from Näf, 1958).

antheridium formation is not observed to be accompanied or preceded by endogenous antheridium production (Döpp, 1950, 1959; Näf, 1961). An important aspect of antheridiogen action could be a prolongation of the period of sensitivity, as was demonstrated by Näf (1962b) in *Onoclea*. This becomes evident when one considers that normally a population of fern gametophytes is not developmentally synchronous. The most rapidly developing ones produce archegonia without, or with little prior (spontaneous) antheridia formation. The production of antheridiogen by these individual may lead to a prolonged antheridia formation in younger, slower developing, and still sensitive ones, so that within a population sexual reproduction is promoted by the simultaneous occurrence of archegonia and antheridia (Näf, 1958; Näf *et al.*, 1975; Döpp, 1962).

Döpp (1959) advocated the concept that the meristem of insensitive prothalli produces an inhibitor which blocks the action of antheridiogen. The postulated inhibitor was thought to be non-diffusible but capable of moving from cell to cell. This was based on the observation that antheridia are rapidly formed when the meristem is removed. A similar phenomenon was observed by Näf (1961), who suggested that during the sensitive stage of growth, antheridiogen acts by reversing the inhibitory effect of actively

138 Sexual Interactions in Plants

dividing cells in the anterior region of the prothallus (cf. also Näf *et al.*, 1975). This may apply to the objects in which the promoting effect of meristem dissection was observed, namely *Pteridium* and *Onoclea*, but not to *Anemia*, in which this treatment was not successful (Schraudolf, 1966b).

POSSIBLE INTERACTION BETWEEN LIGHT AND ANTHERIDIOGEN ACTIONS

Light has a marked influence on antheridium formation in fern gameto-phytes. *Polypodium crassifolium* and *Onoclea sensibilis* fail to form antheridia in the light, but these organs arise spontaneously upon exposure to darkness or near-darkness, respectively (Näf *et al.*, 1974). Antheridiogen A_{An} induces antheridia in *Anemia phyllitidis* to a concentration many times lower in darkness than in the light, as is illustrated in Table XXIII (Näf, 1966, 1967). The possibility that light inactivates the inducing factor was ruled out by the demonstration that the compound, when added to inoculated culture medium, and exposed to routine culture conditions in the light or in the dark for 20 days, could be recovered almost quantitatively, as is shown in Table XXIV. Apart from the fact that this experiment shows that antheridiogen is not, or hardly, metabol-

TABLE XXIII. Antheridium formation in light-and dark-grown *Anemia* prothalli exposed to antheridiogen continuously, starting at the spore stage (from Näf, 1967, modified)

Anemia medium, dilution	Numbers of antheridium-bearing prothalli at intervals (days) following spore inoculation[a]					
	Light-grown prothalli			Dark-grown prothalli		
	3	5	7	3	5	6
1/10	0·0	10·0	27·5	0·0	8·5	15·5
1/30	0·0	6·5	23·5			
1/100		0·0	6·0	0·0	10·5	18·0
1/300		0·0	0·0			
1/1000			0·0	0·0	10·0	14·0
1/10 000				0·0	5·0	10·5
1/30 000				0·0	2·0	8·0

[a] Values are averages of antheridium-bearing individuals in each of two samples of 30 prothalli.

TABLE XXIV. Stability of factor(s) inducing antheridia in culture medium exposed to various conditions for 20 days (from Näf, 1966)[a]
Antheridia-inducing activity in light[b]

Dilution	Anemia medium added at 1/10 full strength					Anemia medium added at 1/100 full strength				
	1	2	3	4	5	1	2	3	4	5
1/3	—	—	—	—	—	18·0	11·5	12·5	7·0	6·0
1/10	—	—	—	—	—	8·5	8·0	8·5	12·5	8·5
1/30	27·5	25·0	22·0	24·5	23·0	0·5	0·0	1·5	1·5	0·0
1/100	15·0	15·0	14·0	12·5	10·5	0·0	0·0	0·0	0·0	0·0
1/300	4·0	5·0	1·5	1·5	2·0	—	—	—	—	—
1/1000	0·0	0·0	0·0	0·0	0·0	—	—	—	—	—
1/3000	—	—	—	—	—	—	—	—	—	—
1/10 000	—	—	—	—	—	—	—	—	—	—
1/30 000	—	—	—	—	—	—	—	—	—	—

[a] Condition 1: Culture medium not inoculated; stored in frozen state and darkness. 2: Culture medium inoculated with Anemia spores exposed to routine conditions of culture in the light. 3: Idem, but culture medium not inoculated. 4: Culture medium inoculated, exposed to darkness but otherwise routine culture conditions. 5: Idem, but culture medium not inoculated.

[b] Values are averages of antheridium-bearing individuals in each of two samples of 30 prothalli.

ized (with the assumption that no "extra" antheridiogen is produced by the treated prothalli), the conclusion must be drawn that light diminishes the competence of prothalli to respond to the active factor (Näf, 1966).

The difference in sensitivity between light- and dark-grown prothalli is particularly illustrated by some experiments with *Onoclea* (Näf *et al.*, 1974). When prothalli of this species, grown for 7 days in the light, are exposed to near-darkness, on a sucrose-containing medium, antheridia arise spontaneously after 15–22 days. Once started, antheridium formation continues indefinitely. When the medium is supplemented with antheridiogen A_{Pt}, sexual morphogenesis is observed already within 4 days, as is illustrated in Fig. 36. A longer period of near-darkness before application of A_{Pt} results in a shorter lag period. In contrast, light-grown *Onoclea* prothalli do not form any antheridia spontaneously and their sensitivity to A_{Pt} disappears completely after 7–11 days of cultivation.

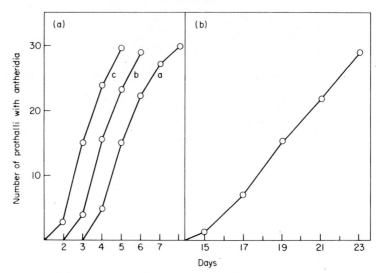

FIG. 36. Time relationships of antheridium formation in near-darkness. Values are averages of antheridium-bearing prothalli in each of four samples of 30 prothalla (a) Lag periods between exposure to A_{Pt} and appearance of antheridium initilis. in near-darkness: a, exposed to A_{Pt} and near-darkness simultaneously; b, exposed to A_{Pt} 5 days; c, exposed to A_{Pt} 12 days after transfer to near-darkness. (b) Antheridium formation on medium not containing antheridiogen (from Näf *et al.*, 1974).

If prothalli which had been exposed to 7 days of near-darkness were illuminated for a variable period of time, the beneficial effect of near-darkness was partly nullified as is shown in Table XXV.

Apparently, light has a profound effect on the sensitivity of prothalli towards antheridiogen. Since also spontaneous antheridium formation is inhibited by light, one could infer that this is caused by decreased sensitivity towards endogenous antheridiogen.

It seems probable that the effect of light on antheridium formation in *Onoclea* is mediated by the phytochrome system, the inhibiting far-red form of the pigment gradually decaying in darkness. This supposition is based on the fact that the effect of light in this species are very similar in qualitative respect to that in *Polypodium crassifolium*, in which the role of phytochrome in sexual morphogenesis has been verified experimentally (Schraudolf, 1967). In this species red-light interruptions of 5 min at 24-h intervals, applied during the whole dark period, inhibit the formation of antheridia, as is shown in Table XXVI. Red-light inhibition is cancelled by a subsequent irradiation with far-red light. However, the action of

TABLE XXV. The effect of illumination on the sensitivity of *Onoclea* prothalli towards antheridiogen A_{Pt}, after a period of 7 days near-darkness (from Näf *et al.*, 1974)

Experimental treatment	Numbers[b] of antheridium-bearing prothalli at intervals (days) after exposure to A_{Pt}			
	1	2	3	4
A	—	0·00	10·25	28·75
B	0·0	11·50	29·00	—
C	0·0	12·25	28·25	—
D	0·0	9·75	28·50	—
E	0·0	7·00	29·75	—
F	0·0	0·25	13·50	29·00
G	0·0	0·00	12·00	27·75
H	0·0	0·00	0·00	0·00

[a] Exposure to near-darkness on medium not supplemented with sucrose. A: Grown continuously in the light; exposed to A_{Pt} at 7-day stage. B: Precultured for 7 days in the light; then transferred to near-darkness for 7 days; then exposed to A_{Pt} in the light. C–H: Precultured for 7 days in the light; then exposed for 7 days to near-darkness; then exposed to light minus A_{Pt} for variable period of time (C: 3 h; D: 6 h; E: 13 h; F: 24 h; G: 35 h; H: 72 h); then exposed to A_{Pt} in the light.

[b] Averages of antheridium-bearing individuals in each of four samples of 30 prothalli.

TABLE XXVI. Inhibition of antheridia formation by red light and reversion of this effect by subsequent far-red light treatment. The plants were examined 6 days after inoculation (from Schraudolf, 1967)

Light treatment per 24 h darkness	Number of prothalli	Percentage of prothalli containing antheridia
	1138	36
5 min red (R)	916	0
5 min far-red (FR)	882	56
5 min R–5 min FR	955	41
5 min R–5 min FR–5 min R	965	0
5 min R–5 min FR–5 min R–5 min FR	1012	45

an antheridia-inducing hormone has not been demonstrated in this species. More experimental work is clearly needed to disclose the relationship between the effects of light and hormones on sexual morphogenesis in these objects.

11.   Sexual Interactions in Flowering Plants

In higher plants, the haploid gametophytic phase is completely dependent in structural and metabolic respects on the diploid phase, the sporophyte. In the plants which have been subject of most interest, the angiosperms, the female gametophyte is wholly concealed by a complicated diploid structure, the pistil. It is not surprising, therefore, that fertilization in these plants is preceded by interaction between the male and the female individual, in which the diploid *and* the haploid phases take part.

Pollen grains are transported to the top surface of the pistils by the wind, insects or other mediators. In appropriate combinations, this leads to germination and production of the pollen tube. This pollen tube penetrates the stigma and grows through the stylar tissue, enters the ovule and discharges from its tip two non-motile gametes (sperm cells) in the immediate vicinity of the egg cell, which consecutively is fertilized by one of these. The fortuitous way by which pollen arrives at the stigma implies that one or more checkpoints must be passed prior to fertilization which serve to guarantee that the meeting gametes are of the same species. Even more than that, many plants are equipped with mechanisms to prevent inbreeding which would occur most frequently in those cases where male and female organs develop in the same flower and ripen simultaneously. This, and the fact that the pollen tube, on its way to the egg cell, is guided efficiently through the complicated tissue of the female sex organ, implies that there is extensive interaction between the sexual partners prior to fertilization. This interaction will be the subject of this chapter.

Other aspects of sexuality in higher plants, like the induction of flower morphogenesis and sex expression, have been described in several recent textbooks, and will not be discussed here. In particular, Evans's book

"The Induction of Flowering" (1969) and his review article about this subject (1971) are recommended for further reading.

THE MALE GAMETOPHYTE

Pollen development takes place in the pollen sacs in the anther and begins with meiosis. This type of division is induced by exogenous factors transmitted from other parts of the plant and proceeds with a high degree of synchrony (Linskens, 1969). Heslop-Harrison (1972) gives an extensive analysis of meiosis during pollen formation. The inner wall of the anther, the tapetum, develops in a coordinate manner together with pollen formation, its main function being to supply the developing pollen with nutritive material. Particularly, it contributes to the formation of the outer wall of the pollen grain, the exine, which mainly consists of sporopollenin, a polymerization product of β-carotene and carotenoid esters (Brooks and Shaw, 1968). The diploid tapetum is also the origin of the proteins present in the cavities of the exine, which are believed to play a role in cell recognition on arrival of the pollen grain at the stigma. The inner layer of the pollen wall is made up of cellulose and pectin, resembling the primary wall of vegetative cells. It contains many enzymes which are involved in pollen germination and pollen tube growth (Knox and Heslop-Harrison, 1970; Mattsson *et al.*, 1974).

The haploid pollen nucleus divides mitotically to give two nuclei, one of which, the vegetative or tube nucleus, becomes the nucleus of the vegetative cell of the gametophyte. The other, generative, nucleus divides once more, giving rise to a pair of sperm cells. This mitosis takes place either in the pollen grain before pollen dispersal or in the tube after germination or in the ovary. Mature pollen grains are therefore either bi- or trinucleate. The sperm cells are considered as true cells within the pollen, originating after mitosis of the generative nucleus by constriction of the double membrane by which the latter is enclosed. Typical cell walls, made up of carbohydrate, seem to be absent, although in early stages callose is present.

At the time of dispersal, pollen grains are in a state of temporary dormancy, which is governed by hydration. In mature pollen the water content is about 10% in most species, and respiration is extremely low. Upon hydration, respiration and other metabolic processes increase very rapidly.

THE FEMALE GAMETOPHYTE

The ovule can be considered as a female gametangium containing one gamete in the so-called embryo sac, which is enclosed by nucellus tissue and one or two integuments (Fig. 37). In the normal case the embryo sac possesses eight haploid nuclei, formed by three consecutive mitoses. Before fertilization two nuclei (the polar nuclei) fuse, giving the diploid fusion nucleus. This nucleus is bound to fuse with the vegetative male nucleus which results in the first triploid primary endosperm nucleus. Of the other nuclei the egg nucleus and two neighbouring nuclei, the synergids, are surrounded by double membranes and are positioned at

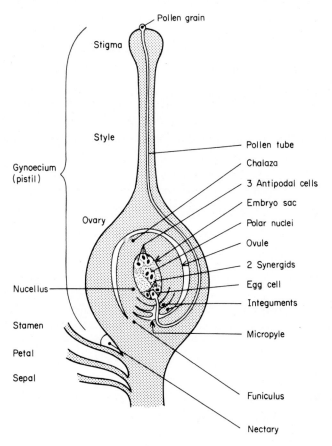

FIG. 37. Vertical section of a flower. The number of ovules per ovary depends on the species (from Linskens, 1969).

one end of the embryo sac, near the opening of the ovule, the micropyle. Three other nuclei are positioned at the other end of the cell.

The most interesting aspect of the style is a zone of tissue connecting the stigma with the ovary, called the stigmatoid or connecting tissue. If the style is hollow, as in the lily, stigmatoid cells line the stylar canal. They probably have a secretory function and might be involved in the orientation of the pollen tube, as will be discussed in a later section.

The stigma is the expanded tip of the style to which pollen grains adhere. Generally, the surface cells of the stigma bear papillae and are covered with a cuticle. In many cases they secrete liquid by which the pollen grains are hydrated and which also contains enzymes involved in pollen germination. Secretion of this liquid stops after fertilization (Linskens, 1969).

Mattsson and coworkers (1974) observed in some eighty angiosperm families that the surface of the stigma papillae bear an external protein coating, overlying the outer layer of the cell wall. This layer, which is called the pellicle, ensheathes the cuticle but without close attachment. It readily forms cracks and fissures, as can be shown by scanning electron microscopy. It can be removed by treatment with proteolytic enzymes. According to these authors, the presence of a proteinaceous and hydrophilic surface layer might facilitate the uptake of water by pollen grains. The water would be derived primarily by capillary force from the underlying cells through discontinuities in the cuticle. Besides this, a hydrated stigma surface could have better adhesive properties.

INCOMPATIBILITY

Fertile pollen grains, brought upon the stigma of a full-grown pistil, do not necessarily accomplish fertilization. In many species mechanisms are present which prevent inter- but also intraspecific fertilization. These incompatibility mechanisms affect pollen germination and pollen tube growth.

Two types of incompatibility mechanisms have been distinguished. The first type is inhibition of pollen germination and pollen tube growth at or in the stigma. This type is called sporophytic incompatibility because the reaction is determined by the genotype of the diploid sporophytes. It is found in Cruciferae and Compositae. That only the stigma is involved in the inhibition of pollen development is demonstrated by the fact that after removal or bypassing of the stigma, pollen grains germinate normally on the wound surface of the stylar tissue and accomplish fertilization in a regular way (Kroh, 1956). The barrier which, according to current concepts, is operative in the stigma is in the first place a specific inhibition of cutinase activity which prevents penetration of the cuticle overlying the stigmatic papillae by the pollen tube. Transfer of pollen from a compatible style to an incompatible style causes normal germination on the latter (Kroh, 1966). This suggests that pollen cutinase is irreversibly activated in the compatible stigma.

More recently, Knox (1973) and Heslop-Harrison et al. (1974) obtained some evidence that callose (β-1,3-glucan) is deposited in the stigma around the location of incompatible pollen grains, which possibly might inhibit the penetration of the pollen tube.

Although the physiological aspects of this type of incompatibility have not fully been elucidated, it is clear that it is based on a specific interaction between the stigmatic surface and the pollen grain which will be discussed in the next section.

In the second type of incompatibility, occurring in a large variety of plants, an interaction between the stylar tissue and the pollen tube is involved. This type is called gametophytic incompatibility because it depends on the genotype of the pollen itself, i.e. the male gametophyte, and the female sporophyte. There is no inhibition of pollen germination, the pollen tube normally penetrates the stigma, but gradually the growth rate diminishes and finally stops altogether. This growth inhibition is accompanied by malformations of the tube, a general disturbance of carbohydrate metabolism, resulting in secondary cellulose and callose appositions near the tip and abnormal behaviour of the nuclei.

POLLEN–STIGMA INTERACTIONS

Since the discovery by Tsinger and Petrovskaya-Baranova (1961) it is generally observed that pollen very rapidly releases proteinaceous material upon arrival at the stigma. Stanley and Linskens (1965) have reported, for example, that in a suspension of *Petunia* pollen proteins are detected within 15 min after preparing the suspension, although tube growth becomes apparent only after half an hour at 26°C. Mäkinen and Brewbaker (1967) have shown that, among others, esterases, catalases, amylases and phosphatases diffuse from *Oenothera* pollen within a few minutes in salt solutions which inhibit germination. Knox and Heslop-Harrison have suggested that most of the enzymes secreted were derived from the wall, particularly the intine, where they were shown to be present by cyto-chemical and immunological techniques. They also have cited older work (Green, 1894, loc cit.) inferring that the biological role of the rapidly released hydrolytic enzymes might be to stimulate pollen germination and to sustain pollen tube growth by dissolving part of the stigma. This empha-sizes the saprophytic nature of the male gametophyte and the concurrent nutritional function of the stigma. Interestingly, the spores of two ferns which germinate to give autotrophic gametophytes do not release hydrolytic enzymes (Knox, 1971; Knox and Heslop-Harrison, 1970, 1971a).

Roggen and Stanley (1969), on the other hand, have hypothesized that hydrolysing enzymes affect the plasticity of the wall at the tip of the pollen tubes, and so affect the growth rate of the tube. The rationale behind this is the well-established fact that pollen tube growth is restricted to the tip, which must therefore be sufficiently plastic to allow extension. Roggen and Stanley have tested this hypothesis by investigating the effect of β-1,4-glucanase, pectinase and β-1,3-glucanase on pollen tube elongation. They found that β-1,3-glucanase stimulated pollen germination, whereas β-1,4-glucanase and pectinase at 0·1–0·25 mg/ml increased tube length by some 25% if added to 1-h-old tubes. At higher concentrations the added enzymes inhibited tube growth. These results suggest that part of the enzymes released by pollen might possibly be involved in tube growth, although it must be added that these authors also found an inhibitory effect of other enzymes released by pollen, notably, proteinase, pectin esterase, acid phosphatase and α-amylase. Possibly these enzymes play a role in releasing substances from the style that are required for tube growth. A detailed study of pollen germination and tube growth in defined media after prior elimination of these pollen-wall enzymes, e.g. by pre-incubating the pollen in a salt solution might yield more evidence with regard to these enzymes.

Several authors, like Stanley and Linskens (1965), have suggested that proteins, released by pollen grains, might comprise not only hydrolytic enzymes, but also proteins involved in the control of fertilization. Knox and Heslop-Harrison (1971b), initially, did not find any difference in the dissipation of intine-held proteins on the surface of the stigma in compatible and incompatible matings. But the suggestion obtained sup-port from experiments by Knox et al. (1972) with the interspecific cross-incompatibility system in poplars. *Populus deltoides* and *P. alba* are incompatible with respect to each other; mating between these two species does not lead to progeny because pollen germination is inhibited at the stigmatic surface. They can, however, be perfectly selfed. When pollen of *P. alba* was mixed with non-viable, γ-irradiated pollen of *P. detoides* and used for pollination of *P. deltoides,* a hybrid progeny was obtained, indicating that the fertilization barrier was relieved by the admixture of non-viable pollen. To test if pollen wall proteins are responsible for this effect, pollen of *P. deltoides* were extracted with 1% NaCl solution for 2 h at $5°C$ with constant agitation. The extract was dialysed against water for 15 h to remove the salt and other low molecular-weight substances, and lyophilized. The residue was thoroughly mixed in equal volumes with *P. alba* pollen and applied to stigmas of *P. deltoides.* The seed set was approximately 8% of the control (selfed *P. deltoides* without treatment), which is low but significant. All progeny was hybrid. A possible explanation for the low fertilization percentage is the fact that unnaturally high amounts of extract were applied together with *P. alba* pollen. Several components of the protein mixture released by pollen are inhibitory to pollen-tube growth in high concentration, as was discussed above (Roggen and Stanley, 1969). Nevertheless, it seems evident from this experiment that pollen proteins play an important role in the recognition process between pollen and stigma, because an extract of compatible pollen can overcome the fertilization barrier in an incompatible mating. Unfortunately, the converse experiment was not described, so we do not know what the effect is of an extract of incompatible pollen in a compatible mating.

The question can be asked whether similar phenomena are evoked in typical intraspecific self-incompatible systems. A positive answer has been provided by the experiments of Heslop-Harrison *et al.* (1974) with Cruciferae. These authors studied two species of *Iberis, I. semperflorens* and *I. sempervirens,* which typically show self-incompatibility at the stigmatic surface level. To test if isolated protein fractions from pollen walls have a function in cell recognition, a rapid assay was developed using the fact that incompatible pollination in these plants gives rise to callose formation in stigma papillae. In a compatible mating pollen tubes develop and penetrate neighbouring stigmatic papillae after dissolution of the cuticle and then grow downwards. The papillae form no internal callose. In incompatible matings, on the other hand, callose plugs are formed in a few hours within the stigma papillae near the pollen grains. If the tubes emerge these also become plugged with callose before they stop growing. This reaction can be observed cytochemically after staining callose specifically. Pollen wall proteins were isolated in two ways. The most rapid method was the pollen-print technique. Pollen to be tested were collected on a strip of Sellotape and this was pressed into contact with an agar film. Under these conditions, mobile wall material diffuses into the gel. The Sellotape was removed after some minutes, carrying with it the adhering pollen grains and leaving prints of wall material in the agar gel. Cubes cut from the gel were applied to the stigmas to test the capacity of the transferred material to induce callose formation. For bulk extractions of wall proteins the method was similar to that described above for the poplar system. It appeared that callose formation was induced when agar cubes with material from incompatible pollen were placed on the stigma, but not when cubes with prints of compatible pollen were used. A similar response was obtained with crude extracts of wall material. Preliminary fractionation by thin-layer gel filtration on Sephadex G150 resulted in a number of protein fractions in the molecular-weight range of 10 000–25 000. All fractions induced some callose formation when brought upon stigmas of an appropriate plant. These results show elegantly that pollen wall proteins are involved in the self-incompatibility reaction in Cruciferae, in as much as callose formation in stigmatic cells is part of it.

In addition, it can be stated that these pollen wall proteins are mainly derived from the outer part of the cell wall, the exine, rather than the intine, because a pollen print, in agar, exposed for no more than 3 min, elicits a full response. In this short period presumably only exine proteins leach out of the pollen. Even more persuasive is the fact that tapetal tissue, applied to the stigma of a flower of the same plant, evokes the typical callose-forming reaction. Since exine material is derived from the tapetum, exclusively, during pollen formation (see page 144), this proves that the active proteins isolated from pollen are at least partly localized in the exine (cf. also Heslop-Harrison et al., 1973).

In summary, these results support the concept that wall proteins in pollen are part of the incompatibility system in a number of higher plant families. These proteins presumably interact with components of the stigmatic surface, which might be located in the proteinaceous pellicle overlaying the cuticle (Mattsson et al., 1974). The nature of this interaction remains to be investigated, as is also the way in which callose formation is started. In addition, other aspects of sporophytic incompatibility, like the cutin esterase activation, mentioned in an earlier section, must still be elucidated.

STYLE–POLLEN TUBE INTERACTIONS

Although pollen tubes of some species can be grown *in vitro* for considerable periods of time without exogenous food supply, there are several
indications that they derive nutrients from the surrounding stylar tissue.
Electron microscopic and autoradiographic studies have shown that pollen
tubes during their growth hydrolyse part of the pectin-containing intercellular material, and absorb the products. Kroh and coworkers (1970)
demonstrated that [^{14}C]myo-inositol, a precursor of pectic substances,
when fed to *Lilium longiflorum* pistils prior to pollination, was incorporated in the pollen cell wall. However, part of the uptake of nutrient
materials might occur via the exudate of the stigmatoid cells. Rosen
(1971) demonstrated that ^{3}H-labelled myo-inositol, fed to cut pistils,
accumulated dramatically at the surface of the stigmatoid cells, as well as
in the papillae of the stigma surface and the surrounding exudate (Fig. 38).

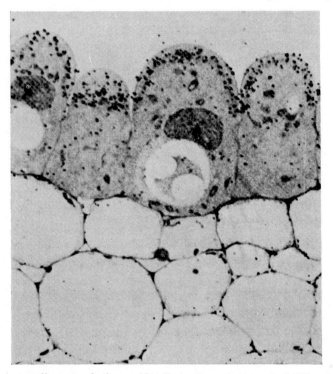

Fig. 38. Autoradiogram of stigmatoid cells in the stylar canal of *Lilium longiflorum*
fed with [^{3}H]myo-insitol. Silver grains along the surface zone of the cells indicate
accumulation of the label ($\times 850$) (from Rosen, 1971).

Kroh *et al.* (1971) obtained labelled pistil exudate by feeding excised pistils with [^{14}C]glucose, and demonstrated that *in vitro* germinated pollen incorporated about 1% of the label. Of this, 38% was recovered from the tube wall fraction. In particular, high-molecular-weight material was used by the tubes for wall synthesis. Taken together, there is much evidence that pollen tubes are nourished by mediation of the secretory stigmatoid cells of the pistil.

It must be added that the stylar stigmatoid tissue frequently becomes secretory only after pollination (Rosen, 1971). It is likely that modifications in the stylar tissue are induced by pollen exudate with the result that nutrients become available for the growing tube (Rosen, 1968; Linskens, 1969).

There is some evidence that specific proteins are involved in the incompatibility interaction betwen style and pollen tube. Linskens (1955) observed that protein mixtures can be isolated from pollinated styles of *Petunia*, the composition of which is dependent on the type of pollen-style combination. From results with tracer experiments it was suggested that one or more components, produced in compatible matings only, are composed of material derived from the male *and* the female individual. On the other hand, the components produced typically in incompatible matings seem to be derived from female material only (Linskens, 1958, 1959). Antibodies against stylar extracts precipitate with pollen extracts of plants with the same incompatibility genotype, and vice versa. When this serological reaction is performed on plants with different incompatibility genotype, the reaction is much weaker or entirely absent. Evidently, incompatibility genes are responsible for the production of substances which are serologically highly specific (Linskens, 1960; Lewis *et al.*, 1967). Conceivably, a recognition process takes place in which components of the male and female individual are involved. Several models have been designed to explain the pollen–style interaction which are all based on an antibody–antigen or enzyme–substrate model. Lewis (1965), for instance, postulates that a particular allele of the incompatibility locus codes for a specific protein pattern, which, when identical in both mates, would associate to form a dimer. This product would directly or indirectly inhibit the growth process of the pollen tube.

Van der Donk (1974a, b) demonstrated that in *Petunia* after pollination in the style quantitative as well as qualitative differences were found in both RNA and protein synthesis between self- and cross-pollinated styles. Following the observation that protein synthesis as a function of time is different in self- and cross-pollinated styles, he investigated RNA synthesis in unpollinated, self-pollinated and cross-pollinated styles. After polyacrylamide gel electrophoresis of labelled RNA a messenger fraction appeared to show the greatest quantitative differences, which was confirmed by injecting these fractions into egg cells of *Xenopus levis* together with [³H]leucine. After incubation, the labelled proteins produced were investigated by polyacrylamide gel electrophoresis in SDS. The protein patterns thus obtained showed distinctive differences, particularly in the low-molecular-weight region. This indicates that gene activity is different in cross- and self-pollinated styles. This might imply that growth inhibition of the pollen tube, in incompatible matings, or, alternatively, growth stimulation in compatible matings, is an inductive effect which becomes effective only some time after penetration of the stigma. For more detailed discussions of this subject the reader is referred to Linskens and Kroh (1967), Rosen (1968) and van der Donk (1974a, b).

CHEMOTROPISM OF POLLEN TUBES

What is the mechanism by which the pollen tube is guided in its growth on its way to the embryo sac? The complicated route it follows leaves little doubt that its growth is directed in one way or another. In the style, pollen tubes grow along the surface of stigmatoid cells (in hollow styles) or they pass through this tissue by intercellular growth. In most plants, the tubes make a 90° turn before entering an ovule. Once through the micropyle, the tubes make again a turn of 90° before penetrating the embryo sac. Most investigators have favoured the idea that orientation of tube growth is by means of chemotropism. This implies that concentration gradients of one or more substances are present in the pistillar tissues, which have a growth-directing influence. Mascarenhas and Machlis (1962a) have reviewed older work on this concept.

Welk and coworkers (1965) focused their attention on the fact that the pollen tube pathway in lily is entirely lined with stigmatoid cells, and tested the idea that these cells excrete a growth-orienting substance. Chemotropic activity was assayed by the so-called depression test designed by Mascarenhas and Machlis (1962b). Three rectangular depressions were made in agar with a mould 3 mm wide, 5 mm long and 3 mm deep, separated by an agar wall 1 mm thick and with an agar floor 2 mm thick. The agar medium consisted of 10% sucrose, 60 p.p.m. boric acid, 100 p.p.m. yeast extract and 1% glucose. The test material was placed in one of the side wells, the other side well serving as a control. Pollen grains were arranged along the side walls of the centre well in the agar. Readings were made when the pollen tubes had almost traversed the agar wall, generally between 4 and 8 h after depositing the grains. A test was indicated as positive when pollen tubes were oriented in a parallel fashion towards the tissue in contrast to random growth on the control side. It must be emphasized that not only orientation of pollen tube growth, but also the mere stimulation of growth might have a "straightening" effect on pollen tubes. For this reason it is necessary that additional tests confirm that it is the concentration gradient of some active factor which is involved. One test could be to apply the pollen several hours after application of the source, in order to observe its effect in the absence of a concentration gradient. Another test would be to demonstrate a change of direction of the pollen tube growth after repositioning the source of the chemotropic substance. Welk and coworkers tested the validity of the depression assay by an experiment in which a pollen tube growth stimulant, gibberellic acid, was assayed for chemotropic activity. In this experiment, more and longer pollen tubes were formed near the test well, but no oriented growth was observed. Also Mascarenhas and Machlis (1964) tested this assay in this way, using sucrose, yeast extract and boric acid as growth stimulants.

With this assay chemotropic activity was demonstrated in exudates of stigmatoid tissue of *Lilium leucanthum* and *L. regale* along the pathway from stigma to ovule. The activity appeared to be dependent on the developmental stage of the tissue, the stigma being active 3–5 days before anthesis, the style 2–3 days and the ovules only 2 days after anthesis. Presumably the factor responsible for orientation was secreted by stigmatoid cells because very small blocks of tissue only evoked a response if they contained stigmatoid cells. Thus the conclusion seems warranted that it is the stigmatoid tissue which exerts a dominating influence on pollen tube growth. This was confirmed by Rosen (1971) who observed chemotropic activity towards the exudate of the stigma and the stylar canal of *L. longiflorum*.

As to the nature of the hypothetical chemotropic factor the results of a decade-long research turned out to be quite at variance for different species. Schildknecht and Benoni (1963a, b) working with *Oenothera longiflora* and *Narcissus* tried to detect chemotropic factors in aqueous and alcoholic extracts of pistils with the following assay method. Pollen grains were germinated on a gelatine gel with nutrients and the percentage of tubes, growing within an angle of 120° into the direction of a rectangular well containing extract, was determined. Thus random growth in this procedure corresponded to 33·3% Macerated pistils evoked a response of 92% oriented tube growth in the assay, as described above. Aqueous extracts and also mixtures of ninhydrin-positive components and sugars gave similar results. However, in the controls, i.e. in the assay without active tissue or mixtures of active compounds, hardly any tube growth occurred, in contrast to the test containing active material. This suggests that in this assay a *stimulation* rather than an *orientation* of tube growth was measured.

Mixtures of amino acids and sugars appeared not to be active in lily (Rosen, 1961) and *Antirrhinum majus.* In the latter species Mascarenhas and Machlis (1964) showed that calcium might be implicated in directed growth of the pollen tube. The number of tubes growing towards the source of calcium appeared to be roughly proportional to the amount of calcium until saturation was reached. Various calcium salts proved to show activity, whereas salts of Mg, Ba, Sr, Na and K, over a wide concentration range, failed to elicit a response. Boric acid, one of the necessary nutrients for pollen germination and tube growth (Brewbaker and Kwack, 1962), also showed some chemotropic activity, but only in the presence of calcium. Boric acid also enhanced the chemotropic response to calcium.

The levels of calcium in various tissues of the female sex organ of *Antirrhinum majus* appeared to be significantly higher than in vegetative parts. The ovules and the placenta were particularly high in calcium. Also the lower third of the style contained more calcium than the upper part.

These encouraging results were set back, however, when Mascarenhas in 1966 by a cytochemical method found that stigmatoid tissue in this plant was lower in calcium than the surrounding tissue. This could not be reconciled with the fact that pollen tubes grow exclusively through this tissue. Neither could a concentration gradient in it be observed. Thus, the spectacular effect of calcium on tube growth *in vitro* was not paralleled by the demonstration of specific and local concentration gradients of this element *in vivo*. In addition, the action of calcium in the depression test on pollen tubes of *Antirrhinum majus* was not reproduced with pollen tubes of lily (Rosen, 1964) and *Oenothera* (Glenck *et al.,* 1973).

What is the reason of this failure to elucidate the mechanism of pollen growth orientation? At any rate, this work has been severely hampered by the fact that no reliable bioassay was available to test orientation of pollen tube growth. According to Rosen (1968), an adequate bioassay must: (a) measure activity as a function of the weight of material applied (tissue or purified fractions); (b) distinguish between chemotropism and physical effects (e.g. thigmotropism); (c) distinguish between non-chemotropic growth stimulation and true chemotropism; and (d) permit a quantitative estimation of chemotropic activity. None of the assays used hitherto fulfils all criteria. A more extensive discussion of chemotropism is given on page 21 ff.

Another possibility, of course, to explain the inability to demonstrate a concentration gradient of a chemotropic factor, is that it does not exist. This possibility was envisaged by Mascarenhas (1973). He remarked that any discussion of the mechanism of action of chemotropic factors must take into consideration the special nature of the growth of a pollen tube. This growth, as defined as wall extension, is restricted to a zone which extends back to no more than a few micrometres from the tip. This region is rich in protein and polysaccharide and contains numerous vesicles which arise from the Golgi apparatus and contribute to the formation of new wall material and plasma membrane at the tip of the tubes (cf. Rosen, 1968, for review; and page 24). Cell wall synthesis is thus visualized as the result of a secretory process, involving carbohydrate precursors and enzymes, which concentrates at the tip of the pollen tube. For a change in the direction of growth all that theoretically is needed is a small shift of the growing zone away from the long axis in the tube. This shift could be effected by a concentration gradient of a tropic factor or by the asymmetrical distribution of substances required for cell wall metabolism. Considering the fact that stigmatoid cells most probably contribute cell wall material to the growing pollen tube (as was described on page 151), one could imagine that a "trail" of agents necessary for tip growth could lead a tube through the appropriate tissue towards the embryo sac. This, of course, would not be chemotropism in the sense of a growth orientation under the influence of a concentration gradient. If this hypothesis of Mascarenhas is correct, there would be no necessity for a continuously increasing concentration gradient of a tropic substance from stigma to embryo sac. The only requirement would be the presence of a limiting growth factor along the pollen tube pathway, or the presence of an enzyme system activated by the pollen tube and providing such a factor (cf. page 152). The activity of calcium and boron could possibly be fitted in this hypothesis, because these elements have been suggested to play a role in pectin synthesis (Kwack, 1964; Stanley and Loewus, 1964).

Mascarenhas derives some support for this hypothesis from experiments performed by Iwanami (1959; cited by Mascarenhas, 1973), who demonstrated that pollen tubes introduced in a lower portion of the hollow lily style grow in about equal numbers towards the stigma and the ovary, along the stigmatoid cells. This suggests that it is not a concentration gradient that is necessary for oriented tube growth, but only the presence of stigmatoid cells which by secretion of factors required for growth would orient the pollen tubes.

F

References

Adler, J. (1973). A method for measuring chemotaxis and use of the method to determine optimum conditions for chemotaxis. *Journal of General Microbiology* **74**, 77–91.

Adler, J., Hazelbauer, G. L. and Dahl, M. M. (1973). Chemotaxis towards sugars in *Escherichia coli*. *Journal of Bacteriology* **115**, 824–847.

Ahmad, M. (1965). Incompatibility in yeasts. *In* "Incompatibility in Fungi" (K. Esser and J. R. Raper, eds), pp. 13–23. Springer-Verlag, Berlin, Heidelberg, New York.

Allison, A. C., Davies, P. and de Petris, S. (1971). Role of contractile microfilaments in macrophage movement and endocytosis. *Nature, New Biology* **232**, 153–155.

Altman, P. L. and Dittmer, D. S., eds (1973). "Biology Data Book," 2nd edn, vol. 2, pp. 661–667. Federation of the American Societies for Experimental Biology, Bethesda, Md.

Amrhein, N. and Filner, P. (1973). Adenosine $3':5'$-cyclic monophosphate in *Chlamydomonas reinhardtii*. *Proceedings of the National Academy of Sciences of the United States of America* **70**, 1099–1103.

Arsenault, G. P., Biemann, K., Barksdale, A. W. and McMorris, T. C. (1968). The structure of antheridiol, a sex hormone in *Achlya bisexualis*. *Journal of the American Chemical Society* **90**, 5635–5636.

Austin, D. J., Bu'Lock, J. D. and Gooday, G. W. (1960a). Trisporic acids: sexual hormones from *Mucor mucedo* and *Blakeslea trispora*. *Nature* **223**, 1178–1179.

Austin, D. G., Bu'Lock, J. D. and Winstanley, D. J. (1969b). Trisporic acid biosynthesis and carotenogenesis in *Blakeslea trispora*. *Biochemical Journal* **113**, 34P.

Austin, D. J., Bu'Lock, J. D. and Drake, D. (1970). The biosynthesis of trisporic acids from β-carotene via retinal and trisporol. *Experentia* **26**, 348–349.

Balsamo, J. and Lilien, J. (1974). Functional identification of three components which mediate tissue-type specific embryonic cell abhesion. *Nature* **251**, 522–524.

Banbury, G. H. (1954). Processes controlling zygophore formation and zygotropism in *Mucor mucedo* Brefeld. *Nature* **173**, 499–500.

Bandoni, R. J. (1963). Conjugation in *Tremella mesenterica*. *Canadian Journal of Botany* **41**, 467–474.

Bandoni, R. J. (1965). Secondary control of conjugation in *Tremella mesenterica*. *Canadian Journal of Botany* **43**, 627–630.

Barksdale, A. W. (1960). Interthallic sexual reactions in *Achlya*, a genus of the aquatic fungi. *American Journal of Botany* **47**, 14–23.

Barksdale, A. W. (1963a). The role of hormone A during sexual conjugation in *Achlya ambisexualis*. *Mycologia* **55**, 627–632.

Barksdale, A. W. (1963b). The uptake of exogenous hormone A by certain strains of *Achlya*. *Mycologia* **55**, 164–171.

Barksdale, A. W. (1965). *Achlya ambisexualis* and a new cross-conjugating species of *Achlya*. *Mycologia* **57**, 493–501.

Barksdale, A. W. (1966). Segregation of sex in the progeny of selfed heterozygote of *Achlya bisexualis*. *Mycologia* **58**, 802–804.

Barksdale, A. W. (1969). Sexual hormones of *Achlya* and other fungi. *Science* **166**, 831–837.

Barksdale, A. W. (1970). Nutrition and antheridiol-induced branching in *Achlya ambisexualis*. *Mycologia* **62**, 411–420.

Barksdale, A. W. and Lasure, L. L. (1973). Induction of gametangial phenotypes in *Achlya*. *Bulletin of the Torrey Botanical Club*, **100**, 199–202.

Barksdale, A. W. and Lasure, L. L. (1974). Production of hormone B by *Achlya heterosexualis*. *Applied Microbiology* **28**, 544–546.

Barksdale, A. W., McMorris, T. C., Seshadri, R., Arunachalam, T., Edwards, J. E., Sundeen, J. and Green, D. M. (1974). Response of *Achlya ambisexualis* E87 to the hormone antheridiol and certain other steroids. *Journal of General Microbiology* **82**, 295–299.

Barnett, H. L., Lilly, V. G. and Krause, R. F. (1956). Increased production of carotene by mixed (+) and (−) cultures of *Choanephora cucurbitarum*. *Science* **123**, 141.

Bartnicki-Garcia, S. (1973). Fundamental aspects of hyphal morphogenesis. *Symposia of the Society for General Microbiology* **23**, 245–267.

Bauch, R. (1925). Untersuchungen über die Entwicklungsgeschichte und Sexualphysiologie der *Ustilago bromivora* and *Ustilago grandis*. *Zeitschrift für Botanik* **17**, 129–177.

Ben-Tal, Y. and Varner, J. E. (1974). An early response to gibberellic acid not requiring protein synthesis. *Plant Physiology* **54**, 813–816.

Berg, H. C. (1975). Bacterial behaviour. *Nature* **254**, 389–392.

Berg, H. C. and Brown, D. A. (1972). Chemotaxis in *Escherichia coli* analysed by three-dimensional tracking. *Nature* **239**, 500–504.

Bergman, K., Burke, P. V., Cerda-Olmedo, E., David, C. N., Delbrück, M., Foster, K. W., Goodell, E. W., Heisenberg, M., Meissner, G., Zalokar, M., Dennison, D. S. and Shropshire, W. (1969). *Phycomyces*. *Bacteriological Reviews* **33**, 99–157.

Bhalerao, U. T., Plattner, J. J. and Rapoport, H. (1970). Synthesis of D,L-sirenin and D,L-isosirenin. *Journal of the American Chemical Society* **92**, 3429–3433.

Bilinski, T., Litwinska, J., Zuk, J. and Gajewski, W. (1973). Synchronization of zygote production in *Saccharomyces cerevisiae*. *Journal of General Microbiology* **79**, 285–292.

Bistis, G. (1956). Sexuality in *Ascobolus stercorarius*. I. Morphology of the ascogonium: evidence for a sexual hormonal mechanism. *American Journal of Botany* **43**, 389–394.

Bistis, G. (1957). Sexuality in *Ascobolus stercorarius*. II. Preliminary experiments on various aspects of the sexual process. *American Journal of Botany* **44**, 436–443.

Blakeslee, A. F. (1904). Sexual reproduction in the *Mucorineae*. *Proceedings of the National Academy of Sciences of the United States of America* **40**, 205–319.

Blakeslee, A. F. (1920). Sexuality in *Mucors*. *Science* **51**, 375–382.

Blakeslee, A. F. and Cartledge, J. L. (1927). Sexual dimorphism in Mucorales. II. Interspecific reactions. *Botanical Gazette* **84**, 51–58.

Blakeslee, A. F., Welch, D. S. and Bergner, A. D. (1927). Sexual dimorphism in Mucorales. I. Intraspecific reactions. *Botanical Gazette* **84**, 27–50.

Bonner, J. T. (1973). Hormones in social amoebae. *In* "Humoral Control of Growth and Differentiation" (J. LoBue and A. S. Gordon, eds), vol. 2, pp. 81–98. Academic Press, New York and London.

Brewbaker, J. L. and Kwack, B. H. (1963). The essential role of calcium ions in pollen germination and pollen tube growth. *American Journal of Botany* **50**, 859–865.

Brock, T. D. (1959). Mating reaction in *Hansenula wingei*. Relation of cell surface properties to agglutination. *Journal of Bacteriology* **78**, 59–68.

Brock, T. D. (1965a). Biochemical and cellular changes occurring during conjugation in *Hansenula wingei*. *Journal of Bacteriology* **90**, 1019–1025.

Brock, T. D. (1965b). The purification and characterization of an intracellular sex-specific mannan protein from yeast. *Proceedings of the National Academy of Sciences of the United States of America* **54**, 1104–1112.

Brooks, J. and Shaw, G. (1968). Chemical structure of the exine of pollen walls and a new function for carotenoids in nature. *Nature* **219**, 532–533.

Brown, R. M., Johnson, C. and Bold, M. C. (1968). Electron and phase-contrast microscopy of sexual reproduction in *Chlamydomonas moewusii*. *Journal of Phycology* **4**, 100–120.

Bücking-Throm, E., Duntze, W., Hartwell, L. H. and Manney, T. R. (1973). Reversible arrest of haploid yeast cells at the initiation of DNA synthesis by a diffusible sex factor. *Experimental Cell Research* **76**, 99–110.

Buller, A. H. R. (1933). The formation of hyphal fusions in the mycelium of the higher fungi. *In* "Researches on Fungi", vol. 5, pp. 1–74. Longmans, Green, London.

Bu'Lock, J. D. and Osagie, A. U. (1973). Prenols and ubiquinones in single-strain and mated cultures of *Blakeslea trispora*. *Journal of General Microbiology* **76**, 77–83.

Sexual Interactions in Plants

162

Bu'Lock, J. D. and Winstanley, D. J. (1971). Carotenoid metabolism and sexuality in Mucorales. *Journal of General Microbiology* **68**, xvi–xvii.

Bu'Lock, J. D., Austin, D. J., Snatzke, G. and Kruban, L. (1970). Absolute configuration of trisporic acids and the stereochemistry of cyclization in β-carotene biosynthesis. *Chemical Communications*, 255–256.

Bu'Lock, J. D., Drake, D. and Winstanley, D. J. (1972). Specificity and transformations of the trisporic acid series of fungal sex hormones. *Phytochemistry* **11**, 2011–2018.

Bu'Lock, J. D., Jones, B. E., Quarrie, S. A. and Winskill, N. (1973). The biochemical basis of sexuality in Mucorales. *Naturwissenschaften* **60**, 550–551.

Bu'Lock, J. D., Jones, B. E., Taylor, D., Winskill, N. and Quarrie, S. A. (1974a) Sex hormones in Mucorales. The incorporation of C_{20} and C_{18} precursors into trisporic acids. *Journal of General Micrabiology* **80**, 301–306.

Bu'Lock, J. D., Winskill, N. and Jones, B. E. (1974b). Structures of the mating type specific prohormones of Mucorales. *Chemical Communications*, 708–711.

Burgeff, H. (1924). Untersuchungen über Sexualität und Parasitismus bei Mucorineen. *Botanische Abhandlungen* (K. Goebel, ed.) **4**, 1–135.

Burger, M. M., Lemon, S. M. and Radius, R. (1971). Sponge aggregation. I. Are carbohydrates involved? *Biological Bulletin.* **141**, 380.

Burnett, J. H. (1956). The mating systems in fungi. I. *New Phytologist* **55**, 50–90.

Burnett, J. H. (1968). "Fundamentals of Mycology". Edward Arnold, London.

Butcher, R. W. and Sutherland, E. W. (1962). Adenosine-3′,5′-phosphate in biological materials, *Journal of Biological Chemistry* **237**, 1244–1250.

Caglioti, L., Cainelli, G., Camerino, B., Mondelli, R., Prieto, A., Quilico, A., Salvatori, T. and Selva, A. (1967). The structure of trisporic acid-C acid. *Tetrahedron Suppl.* **7**, 175–187.

Cainelli, G., Grasselli, P. and Selva, A. (1967). Struttura dell'acido trisporico B. *Chimica e l'industria (Milano)* **49**, 628–629.

Calleja, G. B. and Johnson, B. F. (1971). Flocculation in a fission yeast: an initial step in the conjugation process. *Canadian Journal of Microbiology* **17**, 1175–1177.

Campbell, D. A. (1973). Kinetics of the mating specific aggregation in *Saccharomyces cerevisiae*. *Journal of Bacteriology* **116**, 323–330.

Carlile, M. J. and Machlis, L. (1965). The response of male gametes of *Allomyces* to the sexual hormone sirenin. *American Journal of Botany* **52**, 478–483.

Carr, D. J. (ed.) (1972). "Plant Growth Substances". Springer-Verlag, Berlin, Heidelberg and New York.

Chapman, V. J. and Chapman, D. J. (1973). "The Algae", 2nd edn. Macmillan, London.

Ciegler, A. (1965). Microbial carotenogenesis. *Advances in Applied Microbiology* **7**, 1–34.

Cleland, R. (1973). Auxin-induced hydrogen ion excretion from *Avena* coleoptiles. *Proceedings of the National Academy of Sciences of the United States of America* **70**, 3092–3093.

Coleman, A. W. (1962). Sexuality. In "Physiology and Biochemistry of Algae" (R. A. Lewin, ed.), pp. 711–730. Academic Press, New York and London.

Conti, S. F. and Brock, T. D. (1965). Electron microscopy of cell fusion in conjugating *Hansenula wingei*. *Journal of Bacteriology* **90**, 524–533.

Cook, A. H. and Elvidge, J. A. (1951). Fertilization in the Fucaceae: investigations of the nature of the chemotactic substance produced by the eggs of *Fucus serratus* and *F. vesiculosus*. *Proceedings of the Royal Society B* **138**, 97–114.

Corey, E. J. and Achiwa, K. (1970). A simple synthetic route to D,L-sirenin. *Tetrahedron Letters* **26**, 2245–2246.

Corey, E. J., Achiwa, K. and Katzenellenbogen, J. A. (1969). Total synthesis of D,L-sirenin. *Journal of the American Chemical Society* **91**, 4318–4320.

Crandall, M. A. and Brock, T. D. (1968). Molecular basis of mating in the yeast *Hansenula wingei*. *Bacteriological Reviews* **32**, 139–163.

Crandall, M. A. and Caulton, J. H. (1973). Induction of glycoprotein mating factors in diploid yeast of *Hansenula wingei* by vanadium salts or chelating agents. *Experimental Cell Research* **82**, 159–167.

Crandall, M., Lawrence, L. M. and Saunders, R. M. (1974). Molecular complementarity of yeast glycoprotein mating factors. *Proceedings of the National Academy of Sciences of the United States of America* **71**, 26–29.

Cuatrecasas, P. (1974). Membrane receptors. *Annual Review of Biochemistry* **43**, 169–214.

Cummins, J. E. and Day, A. W. (1973). The cell cycle regulations of mating type alleles in the smut *Ustilago violacea*. *Nature* **245**, 259–261.

Cummins, J. E. and Day, A. W. (1974). Transcription and translation of the sex message in the smut *Ustilago violacea*. II. The effects of inhibitors. *Journal of Cell Science* **16**, 49–62.

Cummins, W. R., Kende, H. and Raschke, K. (1971). Specificity and reversibility of the rapid stomatal response to abscisic acid. *Planta* **99**, 347–351.

Dahlquist, F. W., Lonely, P. and Koshland, D. E. Jr (1972). Quantitative analysis of bacterial migration in chemotaxis. *Nature* **236**, 120–123.

Darden, W. H. (1966). Sexual differentiation in *Volvox aureus*. *Journal of Protozoology* **13**, 239–255.

Darden, W. H. (1973). Hormonal control of sexuality in algae. *In* "Humoral Control of Growth and Differentiation" (J. LoBue and A. S. Gordon, eds), vol. 2, pp. 101–119. Academic Press, London and New York.

Day, A. W. and Cummins, J. E. (1974). Transcription and translation of the sex message in the smut fungus *Ustilago violacea*. I. The effect of ultra-violet light. *Journal of Cell Science* **14**, 451–460.

De Bary, A. (1881). Cited by L. Machlis and E. Rawitscher-Kunkel (1967).

Döpp, W. (1950). Eine die Antheridienbildung bei Farnen fördernde Substanz in den prothallien von *Pteridium aquilinum* (L) Kuhn. *Bericht der Deutschen botanischen Gesellschaft* **63**, 139–147.

Döpp, W. (1959). Uber eine hemmende und eine fördernde Substanz bei der Antheridienbildung in den Prothallien von *Pteridium aquilinum*. *Bericht der Deutschen botanischen Gesellschaft* **72**, 11–24.

Döpp, W. (1962). Weitere Untersuchungen über die Physiologie der Antheridienbildung bei *Pteridium aquilinum*. *Planta* **58**, 483–508.

Dubois-Tylski, Th. (1972). Le cycle de *Closterium moniliferum in vitro*. *Mémoires de la Société botanique de France*, 183–200.

Duntze, W., MacKay, V. and Manney, T. R. (1970). *Saccharomyces cerevisiae*: a diffusible sex factor. *Science* **168**, 1472–1473.

Duntze, W., Stötzler, D., Bücking-Throm, E. and Kalbitzer, S. (1973). Purification and partial characterization of α factor, a mating-type specific inhibitor of cell reproduction from *Saccharomyces cerevisiae*. *European Journal of Biochemistry* **35**, 357–365.

Edwards, J. A., Mills, J. S., Sundeen, J. and Fried, J. H. (1969). The synthesis of the fungal sex hormone antheridiol. *Journal of the American Chemical Society* **91**, 1248–1249.

Edwards, J. A., Schwartz, V., Fajkos, J., Maddox, M. L. and Fried, J. H. (1971). Fungal sex hormones. The synthesis of $(\pm)-7(t),9(t)$-trisporic acid B methyl ester. The stereochemistry at C-9 of the trisporic acids. *Chemical Communications*, 292–293.

Edwards, J. A., Sundeen, J., Salmond, W., Iwadare, T. and Fried, J. H. (1972). A new synthetic route to the fungal sex hormone antheridiol and the determination of its absolute stereochemistry. *Tetrahedron Letters* **9**, 791–794.

Ely, T. H. and Darden, W. H. (1972). Concentration and purification of the male-inducing substance from *Volvox aureus* м5. *Microbios* **5**, 51–56.

Endo, M., Nakanishi, K., Näf, U., McKeon, W. and Walker, R. (1972). Isolation of the antheridiogen of *Anemia phyllitidis*. *Physiologia plantarum* **26**, 183–185.

Esser, K. (1967). Die Verbreitung der Incompatibilität bei Thallophyten. *In* "Handbuch der Pflanzenphysiologie" (W. Ruhland, ed.), vol. **18**, pp. 321–343. Springer-Verlag, Berlin, Heidelberg and New York.

Esser, K. and Kuenen, R. (1965). "Genetik der Pilze". Springer-Verlag, Berlin, Heidelberg, New York.

Evans, M. L. (1974). Rapid responses to plant hormones. *Annual Review of Plant Physiology* **25**, 195–223.

Evans, L. T. (ed.) (1969). "The Induction of Flowering. Some Case Histories". Cornell University Press, Ithaca, N.Y.

Evans, L. T. (1971). Flower induction and the florigen concept. *Annual Review of Plant Physiology* **22**, 365–394.

Feldman, M. and Globerson, A. (1974). Reception of immunogenic signals by lymphocytes. *In* "Current Topics of Developmental Biology" (A. A. Moscona and A. Monroy, eds), vol. 8, pp. 1–40. Academic Press, New York and London.

Finney, D. J. (1964). "Statistical Method in Biological Assay", 2nd edn. Clarks Griffin, London.

Fischer, F. G. and Werner, G. (1955). Eine Analyse des Chemotropismus einiger Pilze, ins besondere der Saprolegniaceae. *Hoppe-Seyler's Zeitschrift für physiologische Chemie* **300**, 211–236.

Förster, H. (1967). Genetik der Geschlechtsbestimmung. *In* "Handbuch der Pflanzenphysiologie" (W. Ruhland, ed.), vol. 18, pp. 31–50. Springer-Verlag, Berlin, Heidelberg and New York.

Förster, H. and Wiese, L. (1954). Gamonwirkungen bei *Chlamydomonas eugametos*. *Zeitschrift für Naturforschung* **9b**, 548–550.

Förster, H. and Wiese, L. (1955). Gamonwirkung bei *Chlamydomonas reinhardi*. *Zeitschrift für Naturforschung* **10b**, 91–92.

Förster, H., Wiese, L. and Braunitzer, G. (1956). Uber das agglutinierend wirkende Gynogamon von *Chlamydomonas eugametos*. *Zeitschrift für Naturforschung* **11b**, 315–317.

Fowell, R. R. (1969). Sporulation and hybridization of yeasts. *In* "The Yeasts" (A. H. Rose and J. S. Harrison, eds), vol. 1, pp. 303–383. Academic Press, London and New York.

Friedmann, I., Colwin, A. L. and Colwin L. H. (1968). Fine structural aspects of fertilization in *Chlamydomonas reinhardi*. *Journal of Cell Science* **3**, 115–128.

Gerisch, G., Beug, H., Malchow, D., Schwarz, H. and Stein, A. von (1974). Receptors for intercellular signals in aggregating cells of the slime mould *Dictyostelium discoideum*. *In* "Biology and Chemistry of Eukaryotic Cell Surfaces" (E. Y. C. Lee and E. E. Smith, eds), pp. 49–66. Academic Press, New York and London.

Glenck, H. O., Wagner, W. and Schimmer, O. (1973). Can Ca^{2+} ions act as chemotropic factor in *Oenothera* fertilization? *In* "Pollen: Development and Physiology" (J. Heslop-Harrison, ed.), pp. 255–261. Butterworths, London.

Gooday, G. W. (1968). Hormonal control of sexual reproduction in *Mucor mucedo*. *New Phytologist* **67**, 815–821.

Gooday, G. W. (1973). Differentiation in the Mucorales. *Symposia of the Society for General Microbiology* **23**, 269–294.

Gooday, G. W., Fawcett, P., Green, D. and Shaw, J. (1973). The formation of fungal sporopollenin in the zygospore wall of *Mucor mucedo*: a role for the sexual carotenogenesis in the Mucorales. *Journal of General Microbiology* **74**, 233–239.

Grieco, P. A. (1969). The total synthesis of D,L-sirenin. *Journal of the American Chemical Society* **91**, 5660–5661.

Halvorson, H. O., Carter, B. L. A. and Tanro, P. (1971). The use of synchronous cultures in yeast to study gene position. *In* "Methods in Enzymology" (L. Grossman and K. Moldave, eds), vol. 21, pp. 462–470. Academic Press, New York and London.

Hartmann, M. (1956). "Die Sexualität". Gustav Fischer Verlag, Stuttgart.

Hartwell, L. H. (1973). Synchronization of haploid yeast cell cycles. A prelude to conjugation. *Experimental Cell Research* **76**, 111–117.

Hartwell, L. H. (1974). *Saccharomyces cerevisiae* cell cycle. *Bacteriological Reviews* **38**, 164–198.

Hassing, G. S., Goldstein, I. J. and Marini, M. (1971). The role of protein carboxyl groups in carbohydrate–concanavalin A interactions. *Biochimica et Biophysica Acta* **243**, 90–97.

Hawker, L. E. (1957). "The Physiology of Reproduction in Fungi". Cambridge University Press.

Hawker, L. E. (1966). Environmental influences on reproduction. *In* "The Fungi, an Advanced Treatise" (G. C. Ainsworth and A. S. Sussman, eds), vol. 2, pp. 435–472. Academic Press, New York and London.

Hawker, L. E. and Beckett, A. (1971). Fine structure and development of the zygospore of *Rhizopus sexualis*. *Philosophical Transactions of the Royal Society, Series B* **263**, 71–100.

Hawker, L. E., Abbott, P. McV. and Gooday, M. A. (1968). Internal changes in hyphae of *Rhizopus sexualis* and *Mucor hiemalis* associated with zygospore formation. *Annals of Botany* **32**, 137–151.

Hazelbauer, G. L. and Adler, J. (1971). Role of the galactose-binding protein in chemotaxis of *Escherichia coli* towards galactose. *Nature, New Biology* **230**, 101–104.

Held, A. A. (1974). Attraction and attachment of zoospores of the parasitic chytrid *Rozella allomycis* in response to host-dependent factors. *Archiv für Mikrobiologie* **95**, 97–114.

Hepden, P. M. and Hawker, L. E. (1961). A volatile substance controlling early stages of zygospore formation in *Rhizopus sexualis*. *Journal of General Microbiology* **24**, 155–164.

Hereford, L. M. and Hartwell, L. H. (1974). Sequential gene function in the initiation of *Saccharomyces cerevisiae* DNA synthesis. *Journal of Molecular Biology* **84**, 445–461.

Herman, A. I. (1971). Sex-specific growth response in yeasts. *Antonie van Leeuwenhoek* **37**, 379–384.

Hertel, R., Thomson, K. and Russo, V. E. A. (1972). *In vitro* auxin binding to particulate cell fractions from corn coleoptiles. *Planta* **107**, 325–340.

Heslop-Harrison, J. (1972). Sexuality of angiosperms. *In* "Plant Physiology, a Treatise" (F. C. Steward, ed.), vol. 6c, pp. 133–291. Academic Press, New York and London.

Heslop-Harrison, J., Heslop-Harrison, Y., Knox, R. B. and Howlett, B. (1973). Pollen wall proteins: "*Gametophytic*" and "*sporophytic*" fractions in the pollen walls of the Malvaceae. *Annals of Botany* **37**, 403–412.

Heslop-Harrison, J., Knox, R. B. and Heslop-Harrison, Y. (1974). Pollen-wall proteins: exine-held fractions associated with the incompatibility response in Cruciferae. *Theoretical Genetics* **44**, 133–137.

Hesseltine, C. W. (1961). Carotenoids in the fungi Mucorales. *Technical Bulletin, United States Department of Agriculture* no. 1245.

Hoffman, L. R. (1960). Cited by L. Machlis and E. Rawitscher-Kunkel (1967).

Hoffman, L. R. (1974). Fertilization in *Oedogonium*. III. Karyogamy. *American Journal of Botany* **61**, 1076–1090.

Horowitz, D. K. and Russell, P. J. (1974). Hormone-induced differentiation of antheridial branches in *Achlya ambisexualis*: dependence on ribonucleic acid synthesis. *Canadian Journal of Microbiology* **20**, 977–980.

Isoe, S., Hayase, Y. and Sakan, T. (1971). Sexual hormones of the Mucorales. The synthesis of methyl trisporate B and C. *Tetrahedron Letters* 3691-3694.

Jacob, H. (1962). Technique de synchronization de la formation des zygotes chez la levure *Saccharomyces cerevisiae*. *Comptes rendus des Séances de la Société de biologie* **254**, 3909–3911.

Jaenicke, L. (1972). "Sexuallockstoffe im Pflanzenreich". Vortrag N217 Rheinisch-Westfälischen Akademie der Wissenschaften. Westdeutscher Verlag, Opladen.

Jaenicke, L. (1974). Chemical signal transmission by gamete attractants in brown algae. *In* "Biochemistry of Sensory Functions" (L. Jaenicke, ed.), pp. 307–309. Springer-Verlag, Berlin, Heidelberg and New York.

Jaenicke, L. (1975). Signalstoffe und Chemorezeption bei niederen Pflanzen. *Chemie in unserer Zeit* **9**, 50–59.

Jaenicke, L. and Seferiadis, K. (1975). Die Stereochemie von Fucoserraten, dem Gametenlockstoff der Braunalge *Fucus serratus* L. *Chemische Berichte* **108**, 225–232.

Jaenicke, L., Donike, M. and Akintobi, T. (1971). Sex attractant in a brown alga: chemical structure. *Science* **171**, 815–817.

Jaenicke, L., Akintobi, T. and Marner, F. J. (1973). Ein Beitrag zur Darstellung von Alkyl-cycloheptadienin: Synthese von Ectocarpen und seiner Homologen. *Justus Liebigs Annalen der Chemie* 1252–1262.

Jaenicke, L., Müller, D. G. and Moore, R. E. (1974). Multifidene and aucantene, C_{11} hydrocarbons in the male-attracting essential oil from the gynogametes of *Cutleria multifida* (Smith) Grev. (Phaeophyta). *Journal of the American Chemical Society* **96**, 3324–3325.

Jensen, E. V. and De Sombre, E. R. (1972). Mechanism of action of the female sex hormones. *Annual Review of Biochemistry* **41**, 203–230.

Jensen, E. V., Numata, M., Brecher, P. J. and DeSombre, E. R. (1971). The biochemistry of steroid hormone action. *Biochemical Society Symposia* **32**, 133–159.

Johnson, K. D. and Kende, H. (1971). Hormonal control of lecithin syntheis in barley aleurone cells: regulation of the CDP–choline pathway by gibberellin. *Proceedings of the National Academy of Sciences of the United States of America* **68**, 2674–2677.

Jones, R. L. (1969). Gibberellic acid and the fine structure of barley aleurone cells. I. Changes during the lag phase of α-amylase synthesis. *Planta* **87**, 119–133.

Jones, R. L. (1973). Gibberellins: their physiological role. *Annual Review of Plant Physiology* **24**, 571–598.

Jost, J. P. and Rickenberg, H. V. (1971). Cyclic AMP. *Annual Review of Biochemistry* **40**, 741–774.

Kaldewey, H. and Vardar, Y. (eds) (1972). "Hormonal Regulation in Plant Growth and Development". Verlag Chemie, Weinheim.

Karlson, P. (1973). Sexualpheromone der Schmetterlinge als Modelle chemischer Kommunikation. *Naturwissenschaften* **60**, 113–121.

Kniep, H. (1928). "Die Sexualität der niederen Pflanzen". Jena Verlag von Gustav Fischer.

Kniep, H. (1929). Vererbungserscheinungen bei Pilzen. *Bibliographia genetica* **5**, 371–415.

Knox, R. B. (1971). Pollen wall proteins: localization, enzymic and antigenic activity during development in *Gladiolus* (Iridaceae). *Journal of Cell Science* **9**, 209–237.

Knox, R. B. (1973). Pollen wall proteins: pollen stigma interactions in ragweed and *Cosmos* (Compositae). *Journal of Cell Science* **12**, 421–443.

Knox, R. B. and Heslop-Harrison, J. (1970). Pollen wall proteins: localization and enzymic activity. *Journal of Cell Science* **6**, 1–27.

Knox, R. B. and Heslop-Harrison, J. (1971a). Pollen wall proteins: electron microscopic localization in the intine of *Crocus vernus*. *Journal of Cell Science* **8**, 727–733.

Knox, R. B. and Heslop-Harrison, J. (1971b). Pollen wall proteins: the fate of intine-held antigens on the stigma in compatible and incompatible pollinations by *Phalaris tuberosa*. *Journal of Cell Science* **9**, 239–251.

Knox, R. B., Willing, R. R. and Ashford, A. E. (1972). The role of pollen wall proteins as recognition substances in interspecific incompatibility in poplars. *Nature* **237**, 381–383.

Kochert, G. (1968). Differentiation of reproductive cells in *Volvox carteri*. *Journal of Protozoology* **15**, 438–452.

Kochert, G. and Yates, I. (1974). Purification and partial characterization of a glycoprotein sexual inducer from *Volvox carteri*. *Proceedings of the National Academy of Sciences of the United States of America* **71**, 1211–1214.

Koshland, D. E. Jr (1974). Chemotaxis as a model for sensory systems. *FEBS Letters* **40**, s3–s9.

Kroh, M. (1956). Genetische und entwicklungsphysiologische Untersuchungen über die Selbststerilität von *Raphanus raphanistrum*. *Zeitschrift induktive für Abstammungs- u. Vererbungslehre* **87**, 365–384.

Kroh, M. (1966). Cited by Linskens and Kroh (1967).

Kroh, M., Labarca, C. and Loewus, F. (1971). Use of the pistil for pollen tube wall biosynthesis in *Lilium longiflorum*. In "Pollen: Development and Physiology" (J. Heslop-Harrison, ed.), pp. 273–278. Butterworths, London.

Kroh, M., Miki-Hirosige, H., Rosen, W. and Loewus, F. (1970). Incorporation of label into pollen tube walls from myo-inositol-labelled *Lilium longiflorum* pistils. *Plant Physiology* **45**, 92–94.

Kuhns, W. J., Weinbaum, G., Turner, R. and Burger, M. M. (1973). Aggregation factors of marine sponges. In "Humoral Control of Growth and Differentiation" (J. LoBue and A. S. Gordon, eds), vol. 2, pp. 59–79. Academic Press, New York and London.

Kwack, B. H. (1964). On the role of calcium and other cations in pollen germination and growth. *Botanical Magazine (Tokyo)* **77**, 327–332.

Labavitch, J. M. and Ray, P. M. (1974). Relationship between promotion of xyloglucan metabolism and induction of elongation by indole-acetic acid. *Plant Physiology* **54**, 499–502.

Larsen, A. D. and Sypherd, P. S. (1974). Cyclic adenosine 3',5'-monophosphate and morphogenesis in *Mucor racemosus*. *Journal of Bacteriology* **117**, 432–438.

Larsen, S. H., Reader, R. W., Kort, E. N., Tso, W. W. and Adler, J. (1974). Change in direction of flagellar rotation is the basis of the chemotactic response in *Escherichia coli*. *Nature* **249**, 74–77.

Levi, J. D. (1956). Mating reaction in yeast. *Nature* **177**, 753–754.

Levring, T. (1952). Remarks on the submicroscopical capital structure of eggs and spermatozoids of *Fucus* and related genera. *Physiologia plantarum* **5**, 528–539.

Lewis, D. (1965). A protein dimer hypothesis on incompatibility. In "Genetics Today" (S. J. Geerts, ed.), vol. 3, pp. 657–663. Pergamon Press, Oxford.

Lewis, D., Burrage, S. and Walls, D. (1967). Immunological reactions of single pollen grains, electrophoresis and enzymology of pollen protein exudates. *Journal of Experimental Botany* **18**, 371–378.

Linskens, H. F. (1955). Physiologische Untersuchungen der Pollenschlauch-hemmung selbssteriler Petunien. *Zeitschrift für Botanik* **43**, 1–44.

Linskens, H. F. (1958). Zur Frage der Entstehung der Abwehrkörper bei der Inkompatibilitätsreaktion von *Petunia*. I. *Bericht der Deutschen botanischen Gesellschaft* **71**, 3–10.

Linskens, H. F. (1959). Zur Frage der Entstehung der Abwehrkörper bei der Inkompatibilitätsreaktion von *Petunia*. II. *Bericht der Deutschen botanischen Gesellschaft* **72**, 84–92.

Linskens, H. F. (1960). Zur Frage der Entstehung der Abwehrkörper bei der Inkompatibilitätsreaktion von *Petunia*. III. *Zeitschrift für Botanik* **48**, 126–135.

Linskens, H. F. (1969). Fertilization mechanisms in higher plants. In "Fertilization" (C. B. Metz and A. Monroy, eds), vol. 2, pp. 189–253. Academic Press, New York and London.

Linskens, H. F. and Kroh, M. (1967). Inkompatibilität der Phanerogamen. In "Handbuch der Pflanzenphysiologie" (W. Ruhland, ed.), vol. 18, pp. 506–530. Springer-Verlag, Berlin, Heidelberg and New York.

McCracken, M. D. and Starr, R. C. (1970). Induction and development of reproductive cells in the κ32 strains of *Volvox rousseletii*. *Archiv für Protistenkunde* **112**, 262–282.

Machlis, L. (1958). Evidence for a sexual hormone in *Allomyces*. *Physiologia plantarum* **11**, 181–192.

Machlis, L. (1968). The response of wild-type male gametes of *Allomyces* to sirenin. *Plant Physiology* **43**, 1319–1320.

Machlis, L. (1969a). Zoospore chemotaxis in the watermould *Allomyces*. *Physiologia plantarum* **22**, 126–139.

Machlis, L. (1969b). Fertilization induced chemotaxis in the zygotes of the watermould *Allomyces*. *Physiologia plantarum* **22**, 392–400.

Machlis, L. (1972). The coming of age of sex hormones in plants. *Mycologia* **64**, 235–247.

Machlis, L. (1973a). The chemotactic activity of various sirenins and analogues and the uptake of sirenin by the sperm of *Allomyces*. *Plant Physiology* **52**, 527–531.

Machlis, L. (1973b). Factors affecting the stability and accuracy of the bioassay for the sperm attractant sirenin. *Plant Physiology* **52**, 524–527.

Machlis, L. (1973c). The effects of bacteria on the growth and reproduction of *Oedogonium cardiacum*. *Journal of Phycology* **9**, 342–344.

Machlis, L. and Rawitscher-Kunkel, E. (1967). Mechanisms of gametic approach in plants. *In* "Fertilization" (C. B. Metz and A. Monroy, eds), vol. 1, pp. 117–161. Academic Press, New York and London.

Machlis, L., Hill, G. G. C., Steinback, K. E. and Reed, W. (1974). Some characteristics of the sperm attractant from *Oedogonium cardiacum*. *Journal of Phycology* **10**, 199–204

Machlis, L., Nutting, W. H., Williams, M. W. and Rapoport, H. (1966). Production, isolation and characterization of sirenin. *Biochemistry* **5**, 2147–2152.

Machlis, L., Nutting, W. H. and Rapoport, H. (1968). The structure of sirenin. *Journal of the American Chemical Society* **90**, 1674–1676.

McLean, R. J. and Bosmann, H. B. (1975). Cell–cell interactions: enhancement of glycosyl transferase ectoenzyme systems during *Chlamydomonas* gametic contact. *Proceedings of the National Academy of Sciences of the United States of America* **72**, 310–313.

McLean, R. J. and Brown, R. M. (1974). Cell surface differentiation of *Chlamydomonas* during gametogenesis. *Developmental Biology* **36**, 279–285.

McLean, R. J., Laurendi, C. J. and Brown, R. M. (1974). The relationship of gamone to the mating reaction in *Chlamydomonas moewusii*. *Proceedings of the National Academy of Sciences of the United States of America* **71**, 2610–2613.

MacMillan, J. (1974). Recent aspects of the chemistry and biosynthesis of the gibberellins. *In* "The Chemistry and Biochemistry of Plant Hormones" (V. C. Runeckles, E. Sondheimer and D. C. Walton, eds), vol. 7, pp. 1–19. Academic Press, London and New York.

McMorris, T. C. and Barksdale, A. W. (1967). Isolation of a sex hormone from the water mould *Achlya bisexualis*. *Nature* **215**, 320–321.

McMorris, T. C., Seshadri, R., Weihe, G. R., Arsenault, G. P. and Barksdale, A. W. (1975). Structures of oogoniol-1, -2, and -3, steroidal sex hormones of the water mould *Achlya*. *Journal of the American Chemical Society* **97**, 2544–2545.

Macnab, R. and Koshland, D. E. Jr (1972). The gradient-sensing mechanism in bacterial chemotaxis. *Proceedings of the National Academy of Sciences of the United States of America* **69**, 2509–2512.

Mäkinen, Y. and Brewbaker, J. L. (1967). Isoenzyme polymorphism in flowering plants. I. Diffusion of enzymes out of intact pollen grains. *Physiologia plantarum* **20**. 477–482.

Martz, E. and Steinberg, M. S. (1972). The role of cell–cell contact in "contact" inhibition of cell division. A review and new evidence. *Journal of Cellular and Comparative Physiology* **79**, 189–210.

Mascarenhas, J. P. (1966). The distribution of ionic calcium in the tissues of the gynoecium of *Antirrhinum majus*. *Protoplasma* **62**, 53–58.

Mascarenhas, J. P. (1973). Pollen tube chemotropism. *In* "Behaviour of Micro-organisms" (A. Perez-Miravete, ed.), pp. 62–69. Plenum Press, London and New York.

Mascarenhas, J. P. and Machlis, L. (1962a). The hormonal control of the directional growth of pollen tubes. *Vitamins and Hormones* **20**, 347–371.

Mascarenhas, J. P. and Machlis, L. (1962b). The pollen tube chemotropic factor from *Antirrhinum majus*. *American Journal of Botany* **49**, 482–489.

Mascarenhas, J. P. and Machlis, L. (1964). Chemotropic response of the pollen of *Antirrhinum majus* to calcium. *Plant Physiology* **39**, 70–77.

Matile, P. M., Cortal, M., Wiemken, A. and Frey-Wyssling, A. (1971). Isolation of glucanase-containing particles from budding *Saccharomyces cerevisiae*. *Proceedings of the National Academy of Sciences of the United States of America* **68**, 636–640.

Mattsson, O., Knox, R. B., Heslop-Harrison, J. and Heslop-Harrison, Y. (1974). Protein pellicle of stigmatic papillae as a probable recognition site in incompatibility reactions. *Nature* **247**, 298–300.

Merrell, R. and Glaser, L. (1973) Specific recognition of plasma membranes by embryonic cells. *Proceedings of the National Academy of Science of the United States of America* **70**, 2794–2798.

Mesibov, R., Ordal, G. W. and Adler, J. (1973). The range of attractant concentration for bacterial chemotaxis and the threshold and size of response over this range. *Journal of General Physiology* **62**, 203–223.

Mesland, D. A. M., Huisman, J. G. and van den Ende, H. (1974). Volatile sexual hormones in *Mucor mucedo*. *Journal of General Microbiology* **80**, 111–117.

Metzger, H. (1970). The antigen receptor problem. *Annual Review of Biochemistry* **39**, 889–928.

Miller, J. H. (1968). Fern gametophytes as experimental material. *Botanical Reviews* **34**, 361–440.

Moor, H. (1967). Endoplasmic reticulum as the initiator of bud formation in yeast. *Archiv für Mikrobiologie* **57**, 135–146.

Mori, K. and Matsui, M. (1969). Synthesis of racemic sirenin, a plant sex hormone. *Tetrahedron Letters* **51**, 4435–4438.

Müller, D. G. (1967). Generationswechsel, Kernphasenwechsel und Sexualität der Braunalge *Ectocarpus siliculosus* im Kulturversuch. *Planta* **75**, 39–54.

Müller, D. G. (1968). Versuche zur Karakterisierung eines Sexuallockstoffes bei der Braunalge *Ectocarpus siliculosus*. I. Methoden, Isolierung und gaschromatographischer Nachweis. *Planta* **81**, 160–168.

Müller, D. G. (1972). Chemotaxis in brown algae. *Naturwissenschaften* **59**, 166.

Müller, D. G. and Jaenicke, L. (1973). Fucoserraten, the female sex attractant of *Fucus serratus* L. (*Phaeophyta*). *FEBS Letters* **30**, 137–139.

Mullins, J. T. and Ellis, E. A. (1974). Sexual morphogenesis in *Achlya*: ultrastructural basis for the hormonal induction of antheridial hyphae. *Proceedings of the National Academy of Sciences of the United States of America* **71**, 1347–1350.

Näf, U. (1956). The demonstration of a factor concerned with the initiation of antheridia in polypodiaceous ferns. *Growth* **20**, 91–105.

Näf, U. (1958). On the physiology of antheridium formation in the bracken fern *Pteridium aquilinum* (L.) Kuhn. *Physiologia plantarum* **11**, 728–746.

Näf, U. (1959). Control of antheridium formation in the fern species *Anemia phyllitidis*. *Nature* **184**, 798–800.

Näf, U. (1960). On the control of antheridium formation in the fern *Lygodium japonicum*. *Proceedings of the Society for Experimental Biology and Medicine* **105**, 82–86.

Näf, U. (1961). Mode of action of antheridium-inducing substance in ferns. *Nature* **189**, 900–903.

Näf, U. (1962a). Developmental physiology of lower archegoniates. *Annual Review of Plant Physiology* **13**, 507–532.

Näf, U. (1962b). Loss of sensitivity to the antheridial factor in maturing gametophytes of the fern *Onoclea sensibilis*. *Phyton* **18**, 173–182.

Näf, U. (1965). On antheridial metabolism in the fern species *Onoclea sensibilis*. *Plant Physiology* **40**, 888–890.

Näf, U. (1966). On dark germination and antheridium formation in *Anemia phyllitidis*. *Physiologia plantarum* **19**, 1079–1088.

Näf, U. (1967). On the induction of a phase inhibitory to antheridium formation in the juvenile prothallus of the fern species *Anemia phyllitidis*. *Zeitschrift für Pflanzenphysiologie* **56**, 353–365.

Näf, U. (1968). On the separation and identity of fern antheridiogens. *Plant and Cell Physiology* **9**, 27–33.

Näf, U., Nakanishi, K. and Endo, M. (1975). On the physiology and chemistry of fern antheridiogens. *Botanical Reviews* **41**, 315–59.

Näf, U., Sullivan, J. and Cummins, M. (1969). New antheridiogen from the fern *Onoclea sensibilis*. *Science* **163**, 1357–1358.

Näf, U., Sullivan, J. and Cummins, M. (1974). Fern antheridiogen: cancellation of a light-dependent block to antheridium formation. *Developmental Biology* **40**, 355–365.

Nakanishi, K., Endo, M., Näf, U. and Johnson, L. F. (1971). Structure of the antheridium-inducing factor of the fern *Anemia phyllitidis*. *Journal of the American Chemical Society* **93**, 5579–5581.

Newcomb, E. H. (1969). Plant microtubules. *Annual Review of Plant Physiology* **20**, 253–288.

Nicolson, G. L. (1974). Interactions of lectins with animal cell surfaces. *International Review of Cytology* **39**, 89–190.

Nieuwenhuis, M. and van den Ende, H. (1975). Sex specificity of hormone synthesis in *Mucor mucedo*. *Archiv für Mikrobiologie* **102**, 167–169.

Nolan, R. A. and Bal, A. K. (1974). Cellulase localization in hyphae of *Achlya ambisexualis*. *Journal of Bacteriology* **117**, 840–843.

Nossal, R. and Chen, H. (1973). Effects of chemoattractants on the motility of *Escherichia coli*. *Nature* **244**, 253–254.

Nutting, W. H., Rapoport, H. and Machlis, L. (1968). The structure of sirenin. *Journal of the American Chemical Society* **90**, 6434–6438.

Olive, L. S. (1953). The structure and behaviour of fungus nuclei. *Botanical Review* **19**, 439–586.

Pfeffer, W. (1884). Locomotorische Richtungsbewegungen durch chemischen Reize. *Untersuchungen Botanisches Institut, Tübingen* **1**, 364–482.

Plattner, J. J. and Rapoport, H. (1971). The synthesis of D- and L-sirenin and their absolute configurations. *Journal of the American Chemical Society* **93**, 1758–1761.

Plattner, J. J., Bhalerao, U. T. and Rapoport, H. (1969). Synthesis of D,L-sirenin. *Journal of the American Chemical Society* **91**, 4933.

Plempel, M. (1960). Die zygotropische Reaktion bei Mucorineen. I. *Planta* **55**, 254–258.

Plempel, M. (1962). Die zygotropische Reaktion bei Mucorineen. III. *Planta* **58**, 509–520.

Plempel, M. (1963). Die chemischen Grundlagen der Sexualreaktion bei Zygomyceten. *Planta* **59**, 492–508.

Plempel, M. and Dawid, W. (1961). Die zygotropische Reaktion bei Mucorineen. II. *Planta* **56**, 438–446.

Poon, N. H. and Day, A. W. (1974). "Fimbriae" in the fungus *Ustilago violacea*. *Nature* **250**, 648–649.

Poon, N. H., Martin, J. and Day, A. W. (1974). Conjugation in *Ustilago violacea*. I. Morphology. *Canadian Journal of Microbiology* **20**, 187–191.

Prieto, A., Spalla, C., Bianchi, M. and Biffi, G. (1964). Biosynthesis of β-carotene by strains of Choanephoraceae. *Chemistry and Industry* 551.

Pringle, R. B. (1961). Chemical nature of antheriodiogen A, a specific inducer of the male sex organ in certain fern species. *Science* **133**, 284.

Pringle, R. B., Näf, U. and Braun, A. C. (1960). Purification of a specific inducer of the male sex organ in certain fern species. *Nature* **186**, 1066–1067.

Raper, J. R. (1939). Sexual hormones in *Achlya*. I. Indicative evidence for a hormonal coordinating mechanism. *American Journal of Botany* **26**, 639–650.

Raper, J. R. (1940). Sexual hormones in *Achlya*. II. Distance reactions, conclusive evidence for a hormonal coordinating mechanism. *American Journal of Botany* **27**, 162–173.

Raper, J. R. (1942). Sexual hormones in *Achlya*. V. Hormone A¹, a male secreted augmentor or activator of hormone A. *Proceedings of the National Academy of Sciences of the United States of America* **28**, 509–516.

Raper, J. R. (1950a). Sexual hormones in *Achlya*. VI. The hormones of the A complex. *Proceedings of the National Academy of Sciences of the United States of America* **36**, 524–533.

Raper, J. R. (1950b). Sexual hormones in *Achlya*. VII. The hormonal mechanism in homothallic species. *Botanical Gazette* **112**, 1–24.

Raper, J. R. (1952). Chemical regulation of sexual processes in the thallophytes. *Botanical Review* **18**, 447–545.

Raper, J. R. (1955). Some problems of specificity in the sexuality of plants. *In* "Biological Specificity and Growth" (E. G. Butler, ed.), pp. 119–140. Princeton University Press, N.J.

Raper, J. R. (1959). Sexual versatility and evolutionary processes in fungi. *Mycologia* **51**, 107–124.

Raper, J. R. (1966). Life cycles, basic patterns of sexuality and sexual mechanisms. *In* "The Fungi, an Advanced Treatise" (G. C. Ainsworth and A. S. Sussman, eds), vol. 2, pp. 473–511. Academic Press, New York and London.

Rawitscher-Kunkel, E. and Machlis, L. (1962). The hormonal integration of sexual reproduction in *Oedogonium*. *American Journal of Botany* **49**, 177–183.

Ray, P. M. (1973). Regulation of β-glucan synthetase activity by auxin in pea stem tissue. I. Kinetic aspects. *Plant Physiology* **51**, 601–608.

Ray, P. M. (1974). The biochemistry of the action of indoleacetic acid on plant growth. *In* "The Chemistry and Biochemistry of Plant Hormones" (V. C. Runeckles, E. Sondheimer and D. C. Walton, eds), pp. 93–122. Academic Press, New York and London.

Rayburn, W. R. and Starr, R. C. (1974). Morphology and nutrition of *Pandorina unicocca* sp. nov. *Journal of Phycology* **10**, 42–49.

Reid, I. D. (1974). Properties of conjugation hormones (erogens) from the basidiomycete *Tremella mesenterica*. *Canadian Journal of Botany* **52**, 521–524.

Reschke, T. (1969). Die Gamone aus *Blakeslea trispora*. Zur Struktur der Sexualstoffe aus Mucoraceae. I. *Tetrahedron Letters* **39**, 3435–3439.

Retallack, B. and Butler, R. D. (1973). Reproduction in *Bulbochaete hiloensis* (Nordst.) Tiffany. II. Sexual reproduction. *Archiv für Mikrobiologie* **90**, 343–364.

Ringo, D. L. (1967). Flagellar motion and fine structure of the flagellar apparatus in *Chlamydomonas*. *Journal of Cell Biology* **33**, 543–571.

Robinson, P. M. (1969). Aspects of staling in liquid cultures of fungi. *New Phytologist* **68**, 351–357.

Robinson, P. M. (1973a). Autotropism in fungal spores and hyphae. *Botanical Review* **39**, 367–384.

Robinson, P. M. (1973b). Chemotropism in fungi. *Transactions of the British Mycological Society* **61**, 303–313.

Rodbell, M., Birnbauer, L. and Pohl, S. L. (1970). Adenyl cyclase in fat cells. III. Stimulation by secretin and the effects of trypsin on the receptors for lipolytic hormones. *Journal of Biological Chemistry* **245**, 718–722.

Roggen, H. P. J. R. and Stanley, R. G. (1969). Cell-wall hydrolysing enzymes in wall formation as measured by pollen tube extension. *Planta* **84**, 295–303.

Roseman, S. (1970). The synthesis of complex carbohydrates by multi-glycosyltransferase systems and their potential function in intercellular adhesion. *Chemistry and Physics of Lipids* **5**, 270–297.

Roseman, R. (1974). Complex carbohydrates and intercellular adhesion. *In* "Biology and Chemistry of Eukaryotic Cell Surfaces" (E. Y. C. Lee and E. E. Smith, eds), pp. 317–354. Academic Press, New York and London.

Rosen, W. G. (1961). Studies on the pollen tube chemotropism. *American Journal of Botany* **48**, 889–895.

Rosen, W. G. (1964). Chemotropism and fine structure of pollen tubes. *In* "Pollen Physiology and Fertilization" (H. F. Linskens, ed.), pp. 159–171. North-Holland, Amsterdam.

Rosen, W. G. (1968). Ultrastructure and physiology of pollen. *Annual Review of Plant Physiology* **19**, 435–462.

Rosen, W. G. (1971). Pistil–pollen interactions in *Lilium*. *In* "Pollen development and physiology" (J. Heslop-Harrison, ed.), pp. 239–254. Butterworths, London.

Roth, S. (1973). A molecular model for cell interactions. *Quarterly Review of Biology* **48**, 541–563.

Roth, S. and White, D. (1972). Intercellular contact and cell-surface galactosyltransferase activity. *Proceedings of the National Academy of Sciences of the United States of America* **69**, 485–489.

Roth, S., McGuire, E. J. and Roseman, S. (1971). An assay for intercellular adhesive specificity. *Journal of Cell Biology* **51**, 525–535.

Rubinstein and Light (1973). Indoleacetic-acid-enhanced chloride uptake into coleoptile cells. *Planta*, **110**, 43–56

Sakai, K. and Yanagishima, N. (1972). Mating reaction in *Saccharomyces cerevisiae*. II. Hormonal regulation of agglutinability in *a* type cells. *Archiv für Mikrobiologie* **84**, 191–198.

Satina, S. and Blakeslee, A. F. (1930). Imperfect sexual reactions in homothallic and heterothallic Mucors. *Botanical Gazette* **90**, 299–311.

Schedlbauer, M. D. (1974). Biological specificity of the antheridiogen from *Ceratopteris thalictroides* (L.) Brongn. *Planta* **116**, 39–43.

Schildknecht, H. and Benoni, H. (1963a). Uber die Chemie der Anziehung von Pollenschlauchen durch die Samenanlagen von Oenotheren. *Zeitschrift für Naturforschung* **186**, 45–54.

Schildknecht, H. and Benoni, H. (1963b). Versuche zur Aufklärung des Pollenschlauchchemotropismus von Narcissen. *Zeitschrift für Naturforschung* **186**, 656–661.

Schipper, M. A. A. (1971). Induction of zygospore production in *Mucor saximontensis*, an agamic straing of *Zygorhynchus moelleri*. *Transactions of the British Mycological Society* **56**, 157–159.

Schmeisser, E. T., Baumgartel, D. M. and Howell, S. H. (1973). Gametic differentiation in *Chlamydomonas reinhardi*: cell cycle dependency and rates in attainment of mating competency. *Developmental Biology* **31**, 31–37.

Schraudolf, H. (1964). Relative activity of the gibberellins in the antheridium induction in *Anemia phyllitidis*. *Nature* **201**, 98–99.

Schraudolf, H. (1966a). Die Wirkung von Phytohormonen auf Keimung und Entwicklung von Fernprothallien. IV. Die Wirkung von unterschiedligen Gibberelline und von Allo-gibberellinsäure auf die Auslösung der Antheridienbildung bei *Anemia phyllitidis* L. und einigen Polypodiaceen. *Plant Cell Physiology* **7**, 277–289.

Schraudolf, H. (1966b). Die Wirkung von Phytohormonen auf Keimung und Entwicklung von Fernprothallien. III. Einfluss von Plasmolyse und Extirpation auf die Auslösung der Antheridienbildung durch Gibberelline bei *Anemia phyllitidis*. *Biologisches Zentralblatt* **85**, 349–360.

Schraudolf, H. (1967). Die Steuerung der Antheridienbildung in *Polypodium crassifolium* L. (*Pessopteris crassifolia* Underw. and Maxon) durch Licht. *Planta* **76**, 37–46.

Sebek, O. and Jäger, H. (1964). Biosynthesis of carotenes by *Blakeslea trispora*. *Abstracts of the 148th Meeting of the American Chemical Society* 99.

Sena, E. P., Radin, D. N. and Fogel, S. (1973). Synchronous mating in yeast. *Proceedings of the National Academy of Sciences of the United States of America* **70**, 1373–1377.

Sessoms, A. H. and Huskey, R. J. (1973). Genetic control of development in *Volvox*: isolation and characterization of morphological mutants. *Proceedings of the National Academy of Sciences of the United States of America* **70**, 1335–1338.

Shimoda, C. and Yanagishima, N. (1973). Mating reaction in *Saccharomyces cerevisiae*. IV. Retardation of deoxyribonucleic acid synthesis. *Physiologia plantarum* **29**, 54–59.

Silver, J. C. and Horgen, P. A. (1974). Hormonal regulation of presumptive mRNA in the fungus *Achlya ambisexualis*. *Nature* **294**, 252–254.

Stanley, R. G. and Linskens, H. F. (1965). Protein diffusion from germinating pollen. *Physiologia plantarum* **18**, 47–53.

Stanley, R. G. and Loewus, F. A. (1964). Boron and myo-inositol in pollen pectin biosynthesis. *In* "Pollen Physiology and Fertilization" (H. F. Linskens, ed.), pp. 128–136. North-Holland, Amsterdam.

Starr, R. C. (1969). Structure, reproduction and differentiation in *Volvox carteri* f. *nagariensis* Iyengar, strains HK9 and 10. *Archiv für Protistenkunde* **111**, 204–222.

Starr, R. C. (1970). Control of differentiation in *Volvox*. *Developmental Biology* Suppl. **4**, 59–100.

Starr, R. C. (1971). Sexual reproduction in *Volvox africanus*. *Contributions to Phycology* 59–66.

Starr, R. C. (1972a). Sexual reproduction in *Volvox dissipatrix*. Abstract. *British Phycological Journal* **7**, 279.

Starr, R. C. (1972b). A working model for the control of differentiation during development of the embryo of *Volvox carteri* f. nagariensis. *Soc. Bot. Franç. Mém.* 175–182.

Starr, R. C. and Jaenicke, L. (1974). Purification and characterization of the hormone initiating sexual morphogenesis in *Volvox carteri* f. *nagariensis* Iyengar. *Proceedings of the National Academy of Sciences of the United States of America* **71**, 1050–1054.

Steward, F. C. and Krikorian, A. D. (1971). "Plants, Chemicals and Growth". Academic Press, New York and London.

Sutter, R. P. (1970). Trisporic acid synthesis in *Blakeslea trispora*. *Science* **168**, 1590–1592.

Sutter, R. P. (1975). Mutations affecting sexual development in *Phycomyces blakesleeanus*. *Proceedings of the National Academy of Sciences of the United States of America* **72**, 127–130.

Sutter, R. P. and Rafelson, M. E. (1968). Separation of β-factor from stimulated β-carotene synthesis in mated cultures of *Blakeslea trispora*. *Journal of Bacteriology* **95**, 426–432.

Sutter, R. P., Capage, D. A., Harrison, T. L. and Keen, W. A. (1973). Trisporic acid biosynthesis in separate plus and minus cultures of *Blakeslea trispora*: identification by Mucor assay of two mating-type-specific components. *Journal of Bacteriology* **114**, 1074–1082.

Sutter, R. P., Harrison, T. L. and Galasko, G. (1974). Trisporic acid biosynthesis in *Blakeslea trispora* via mating type-specific precursors. *Journal of Biological Chemistry* **249**, 2282–2284.

Tanada, T. (1972). Antagonism between indoleacetic acid and abscisic acid on a rapid phytochrome-mediated process. *Nature* **236**, 460–461.

Taylor, N. W. and Orton, W. L. (1968). Sexual agglutination in yeast. VII. Significance of the 1·7S component from reduced 5-agglutinin. *Archives of Biochemistry and Biophysics* **126**, 912–921.

Taylor, N. W. and Orton, W. L. (1971). Cooperation among the active binding sites in the sex-specific agglutinin of the yeast *Hansenula wingei*. *Biochemistry* **10**, 2043–2049.

Terenzi, H. F., Flawia, M. M. and Torres, H. N. (1974). A *Neurospora* morphological mutant showing reduced adenylate cyclase activity. *Biochemical and Biophysical Research Communications* **58**, 990–996.

Thomas, D. des S. (1974). Cytochalasin selectively inhibits synthesis of a secretory protein, cellulase, in *Achlya*. *Nature* **249**, 140–141.

Thomas, D. des S. and Mullins, J. T. (1969). Cellulase induction and wall extension in the water mould *Achlya ambisexualis*. *Physiologia plantarum* **22**, 347–353.

Thomas, D. M. and Goodwin, T. W. (1967). Studies on carotenogenesis in *Blakeslea trispora*. I. General observations on synthesis in mated and un-mated strains. *Phytochemistry* **6**, 355–360.

Thomas, D. M., Harris, R. C., Kirk, J. T. O. and Goodwin, T. W. (1967). Studies on carotenogenesis in *Blakeslea trispora*. II. The mode of action of trisporic acid. *Phytochemistry* **6**, 361–366.

Throm, E. and Duntze, W. (1970). Mating type-dependent inhibition of deoxyribonucleic acid synthesis in *Saccharomyces cerevisiae*. *Journal of Bacteriology* **104**, 1388–1390.

Tsinger, N. V. and Petrovskaya-Baranova, T. P. (1961). Cited by R. B. Knox and J. Heslop-Harrison (1970).

Tsubo, Y. (1957). On the mating reaction of a *Chlamydomonas* with special references to clumping and chemotaxis. *Botanical Magazine (Tokyo)* **70**, 327–334.

Tsubo, Y. (1961). Chemotaxis and sexual behaviour in *Chlamydomonas*. *Journal of Protozoology* **8**, 114–121.

Turner, R. S. and Burger, M. M. (1973). Involvement of a carbohydrate group in the active site for surface guided reassociation of animal cells. *Nature* **244**, 509–510.

Tyler, A. (1947). An autoantibody concept of cell structure, growth and differentiation. *Growth* **10**, 7–19.

Uno, I. and Ishikawa, T. (1973). Purification and identification of the fruiting inducing substances in *Coprinus macrorhizus*. *Journal of Bacteriology* **113**, 1240–1249.

Uyama, A. (1972). Chemical biology of sexual factors in the fungi. I. Induction of the gamete initials by the culture extracts in heterothallic

Absidia glauca. Transactions of the Mycological Society of Japan **13**, 66–70.

van de Berg, W. J. and Starr, R. C. (1971). Structure, reproduction and differentiation in *Volvox gigas* and *Volvox powersii*. *Archiv für Protistenkunde* **113**, 195–219.

van den Ende, H. (1967). Sexual factor of the Mucorales. *Nature* **215**, 211–212.

van den Ende, H. (1968). Relationship between sexuality and carotene synthesis in *Blakeslea trispora*. *Journal of Bacteriology* **96**, 1298–1303.

van den Ende, H. and Stegwee, D. (1971). Physiology of sex in Mucorales. *Botanical Review* **37**, 22–36.

van den Ende, H., Werkman, B. A. and van den Briel, M. L. (1972). Trisporic acid synthesis in mated cultures of the fungus *Blakeslea trispora*. *Archiv für Mikrobiologie* **86**, 175–184.

van den Ende, H., Wiechmann, A. H. C. A., Reyngoud, D. J. and Hendriks, T. (1970). Hormonal interactions in *Mucor mucedo* and *Blakeslea trispora*. *Journal of Bacteriology* **101**, 423–428.

van der Donk, J. A. W. M. (1974a). Gene activity and the incompatibility reaction in *Petunia*. *In* "Fertilization in Higher Plants" (H. F. Linskens, ed.), pp. 279–284. Elsevier, Amsterdam.

van der Donk, J. A. W. M. (1974b). Differential synthesis of RNA in self- and cross-pollinated styles of *Petunia hybrida*. *Molecular and General Genetics* **131**, 1–8.

Varner, J. E. (1974). Gibberellin control of a secretory tissue. *In* "The Chemistry and Biochemistry of Plant Hormones" (V. C. Runeckles, E. Sondheimer and D. C. Walton, eds), pp. 123–130. Academic Press, New York and London.

Villar Palasi, C. and Larner, J. (1970). Glycogen metabolism and glycolytic enzymes. *Annual Review of Biochemistry* **39**, 639–672.

Voeller, B. R. (1964a). Gibberellins: their effect on antheridium formation in fern gametophytes. *Science* **143**, 373–375.

Voeller, B. R. (1964b). Antheridiogens in ferns. *In* "Régulateurs naturels de la croissance végétale". *Colloques nationaux du Centre national de la recherche scientifique* **123**, 665–684.

Voeller, B. R. and Weinberg, E. S. (1967). Antheridium induction and the number of sperms per antheridium in *Anemia phyllitidis*. *American Fern Journal* **57**, 107–112.

Weiss, P. (1947). The problem of specificity in growth and development. *Yale Journal of Biology and Medicine* **19**, 235–278.

Welk, M., Millington, W. F. and Rosen, W. G. (1965). Chemotropic activity and the pathway of the pollen tube in lily. *American Journal of Botany* **52**, 774–781.

Werkman, B. A. and van den Ende, H. (1973). Trisporic acid synthesis in

Blakeslea trispora. Interaction between *plus* and *minus* mating types. *Archiv für Mikrobiologie* **90**, 365–374.

Werkman, B. A. and van den Ende, H. (1974). Trisporic acid synthesis in homothallic and heterothallic Mucorales. *Journal of General Microbiology* **82**, 273–278.

White, J. D. and Sung, W. L. (1974). Alkylation of Hageman's ester. Preparation of an intermediate for trisporic acid synthesis. *Journal of Organic Chemistry* **39**, 2323–2328.

Wickerham, L. J. (1956). Influence of agglutination on zygote formation in *Hansenula wingei*, a new species of yeast. *Comptes Rendus des travaux du Laboratoire Carlsberg. Série physiologie* **26**, 423–443.

Wickerham, L. J. (1958). Sexual agglutination of heterothallic yeasts in diverse taxonomic areas. *Science* **128**, 1504–1505.

Wiese, L. (1965). On sexual agglutination and mating-type substances (gamones) in isogamous heterothallic chlamydomonads. I. Evidence of the identity of the gamones with the surface components responsible for sexual flagellar contact. *Journal of Phycology* **1**, 46–54.

Wiese, L. (1969). The algae. *In* "Fertilization. Comparative Morphology, Biochemistry and Immunology" (C. B. Metz and A. Monroy, eds), vol. 2, pp. 135–188. Academic Press, New York and London.

Wiese, L. and Hayward, P. C. (1972). On sexual agglutination and mating type substances in isogamous dioecious chlamydomonads. III. The sensitivity of sex cell contact of various enzymes. *American Journal of Botany* **59**, 530–536.

Wiese, L. and Jones, R. F. (1963). Studies on gamete copulation in heterothallic chlamydomonads. *Journal of Cellular and Comparative Physiology* **61**, 265–274.

Wiese, L. and Metz, C. B. (1969). On the trypsin sensitivity of gamete contact at fertilization as studied with living gametes in *Chlamydomonas*. *Biological Bulletin* **136**, 483–493.

Wiese, L. and Shoemaker, D. W. (1970). On sexual agglutination and mating type substances (gamones) in isogamous heterothallic chlamydomonads. II. The effect of concanavalin A upon the mating reaction. *Biological Bulletin* **138**, 88–95.

Wilkinson, P. C. (1974). Surface and cell membrane activities of leucocyte chemotactic factors. *Nature* **251**, 58–60.

Wurtz, T. and Jockusch, H. (1975). Sexual differentiation in *Mucor*: trisporic acid response mutants and mutants blocked in zygophore development. *Developmental Biology* **43**, 213–220.

Yen, P. H. and Ballou, C. E. (1974a). Structure and immunochemistry of *Hansenula wingei* Y-2340 mannan. *Biochemistry* **13**, 2420–2427.

Yen, P. H. and Ballou, C. E. (1974b). Partial characterization of the sexual agglutination factor from *Hansenula wingei* Y-2340 type 5 cells. *Biochemistry* **13**, 2428–2437.

Zanno, P. R., Endo, M., Nakanishi, K., Näf, U. and Stein, C. (1972). On the structural diversity of fern antheridiogens. *Naturwissenschaften* **11**, 512.

Ziegler, H. (1962). Chemotropismus. *In* "Handbuch für Pflanzenphysiologie" (W. Ruhland, ed.), vol. 17/2, pp. 398–431. Springer-Verlag, Berlin.

Zycha, H., Siepmann, R. and Linneman, G. (1969). "Mucorales". Verlag von J. Cramer, D-3301 Lehre

Index

A

Absidia glauca, 54, 56, 71
Absidia spinosa, 74
abscisic acid, 10, 11
Achlya sp., 3, 4, 6, 10, 24–26, 35–51
A. ambisexualis, 37, 41, 44, 46, 50, 51
A. bisexualis, 40, 50, 51
A. heterosexualis, 46, 50, 51
adenosine 3′,5′-monophosphate, 7, 8
adrenocorticotropin, 7
agglutination, sexual, 2, 14, 15, 79, 86–94, 98–110
aggregation factor of sponge cells, 12
aleurone cells, 10
Allomyces arbuscula, 15, 16, 28–32
A. macrogynus, 27–32
amino acids, as tropic agents, 28, 42, 44
amino acids, role in morphogenesis, 42, 48
anastomoses, 22
androspores, in *Oedogonium*, 122–123
Anemia hirsuta, 133
A. phyllitidis, 132, 133, 138
antheridiogens, 131–142
antheridiol, 24, 40–51
Antirrhinum majus, 155

A

Ascobolus stercoravius, 23, 24
Ascomycetes, 22
aucantene, 128
autotropism, 21, 25
auxins, 9, 10

B

Basidiomycetes, 5, 22
biological assay, 7, 19, 25–26, 29–30, 39, 57–58, 81, 95, 115, 117, 124, 126, 132, 153, 155
Blakeslea trispora, 54, 58–72
Blastocladiella emersonii, 16
boric acid, 154

C

calcium, 155
callose, 147, 149, 150
carotenoids, 58, 60–64, 68, 70
cell cycle, in yeasts, 83–86
cell surface, role of, 11–14, 79, 86–93, 106–109, 146–150
cellular slime moulds, 21
cellulase, 10, 43, 44
cellulose, 43
Ceratopteris thalicroides, 132
chemotaxis, 2, 14, 15–21, 26–34, 109, 122–130